Learn

# Eureka Math®
## Grade 4
## Module 5

**TEKS EDITION**

Great Minds® is the creator of *Eureka Math*®, *Wit & Wisdom*®, *Alexandria Plan*™, and *PhD Science*®.

**Published by Great Minds PBC.**
**greatminds.org**

Copyright © 2021 Great Minds PBC. Except where otherwise noted, this content is published under a limited public license with the Texas Education Agency. Use limited to Non-Commercial educational purposes. For more information, visit https://gm.greatminds.org/texas.

Printed in the USA
1 2 3 4 5 6 7 8 9 10 CCR 25 24 23 22 21

ISBN 978-1-64929-682-5

## Students, families, and educators:

Thank you for being part of the *Eureka Math*® community, where we celebrate the joy, wonder, and thrill of mathematics.

In the *Eureka Math* classroom, new learning is activated through rich experiences and dialogue. The *Learn* book puts in each student's hands the prompts and problem sequences they need to express and consolidate their learning in class.

### *What is in the Learn book?*

**Application Problems:** Problem solving in a real-world context is a daily part of *Eureka Math*. Students build confidence and perseverance as they apply their knowledge in new and varied situations. The curriculum encourages students to use the RDW process—Read the problem, Draw to make sense of the problem, and Write an equation and a solution. Teachers facilitate as students share their work and explain their solution strategies to one another.

**Problem Sets:** A carefully sequenced Problem Set provides an in-class opportunity for independent work, with multiple entry points for differentiation. Teachers can use the Preparation and Customization process to select "Must Do" problems for each student. Some students will complete more problems than others; what is important is that all students have a 10-minute period to immediately exercise what they've learned, with light support from their teacher.

Students bring the Problem Set with them to the culminating point of each lesson: the Student Debrief. Here, students reflect with their peers and their teacher, articulating and consolidating what they wondered, noticed, and learned that day.

**Exit Tickets:** Students show their teacher what they know through their work on the daily Exit Ticket. This check for understanding provides the teacher with valuable real-time evidence of the efficacy of that day's instruction, giving critical insight into where to focus next.

**Templates:** From time to time, the Application Problem, Problem Set, or other classroom activity requires that students have their own copy of a picture, reusable model, or data set. Each of these templates is provided with the first lesson that requires it.

### *Where can I learn more about* Eureka Math *resources?*

The Great Minds® team is committed to supporting students, families, and educators with an ever-growing library of resources, available at eureka-math.org. The website also offers inspiring stories of success in the *Eureka Math* community. Share your insights and accomplishments with fellow users by becoming a *Eureka Math* Champion.

Best wishes for a year filled with aha moments!

Jill Diniz
Director of Mathematics
Great Minds

# Learn ◆ Practice ◆ Succeed

*Eureka Math*® student materials for *A Story of Units*® (K–5) are available in the *Learn, Practice, Succeed* trio. This series supports differentiation and remediation while keeping student materials organized and accessible. Educators will find that the *Learn, Practice,* and *Succeed* series also offers coherent—and therefore, more effective—resources for Response to Intervention (RTI), extra practice, and summer learning.

## Learn

*Eureka Math Learn* serves as a student's in-class companion where they show their thinking, share what they know, and watch their knowledge build every day. *Learn* assembles the daily classwork—Application Problems, Exit Tickets, Problem Sets, templates—in an easily stored and navigated volume.

## Practice

Each *Eureka Math* lesson begins with a series of energetic, joyous fluency activities, including those found in *Eureka Math Practice.* Students who are fluent in their math facts can master more material more deeply. With *Practice,* students build competence in newly acquired skills and reinforce previous learning in preparation for the next lesson.

Together, *Learn* and *Practice* provide all the print materials students will use for their core math instruction.

## Succeed

*Eureka Math Succeed* enables students to work individually toward mastery. These additional problem sets align lesson by lesson with classroom instruction, making them ideal for use as homework or extra practice. Each problem set is accompanied by a Homework Helper, a set of worked examples that illustrate how to solve similar problems.

Teachers and tutors can use *Succeed* books from prior grade levels as curriculum-consistent tools for filling gaps in foundational knowledge. Students will thrive and progress more quickly as familiar models facilitate connections to their current grade-level content.

# The Read–Draw–Write Process

The *Eureka Math* curriculum supports students as they problem-solve by using a simple, repeatable process introduced by the teacher. The Read–Draw–Write (RDW) process calls for students to

1. Read the problem.
2. Draw and label.
3. Write an equation.
4. Write a word sentence (statement).

Educators are encouraged to scaffold the process by interjecting questions such as

- What do you see?
- Can you draw something?
- What conclusions can you make from your drawing?

The more students participate in reasoning through problems with this systematic, open approach, the more they internalize the thought process and apply it instinctively for years to come.

# Contents

## Module 5: Fraction Equivalence, Ordering, and Operations

**Topic A: Decomposition and Fraction Equivalence**

Lesson 1 .................................................................... 1

Lesson 2 .................................................................... 9

Lesson 3 .................................................................... 15

Lesson 4 .................................................................... 23

Lesson 5 .................................................................... 29

**Topic B: Fraction Equivalence Using Multiplication and Division**

Lesson 6 .................................................................... 37

Lesson 7 .................................................................... 45

Lesson 8 .................................................................... 51

Lesson 9 .................................................................... 59

Lesson 10 ................................................................... 67

**Topic C: Fraction Comparison**

Lesson 11 ................................................................... 75

Lesson 12 ................................................................... 83

Lesson 13 ................................................................... 91

Lesson 14 ................................................................... 99

**Topic D: Fraction Addition and Subtraction**

Lesson 15 ................................................................... 107

Lesson 16 ................................................................... 115

Lesson 17 ................................................................... 121

Lesson 18 ................................................................... 127

**Topic E: Extending Fraction Equivalence to Fractions Greater Than 1**

Lesson 19 . . . . . . . . . . . . . . . . . . . . . . . . . . . . . . . . . . . . . . . . . . . . . . . . . . . . . . . . . . . . . . . . . . 133

Lesson 20 . . . . . . . . . . . . . . . . . . . . . . . . . . . . . . . . . . . . . . . . . . . . . . . . . . . . . . . . . . . . . . . . . . 139

Lesson 21 . . . . . . . . . . . . . . . . . . . . . . . . . . . . . . . . . . . . . . . . . . . . . . . . . . . . . . . . . . . . . . . . . . 145

Lesson 22 . . . . . . . . . . . . . . . . . . . . . . . . . . . . . . . . . . . . . . . . . . . . . . . . . . . . . . . . . . . . . . . . . . 151

Lesson 23 . . . . . . . . . . . . . . . . . . . . . . . . . . . . . . . . . . . . . . . . . . . . . . . . . . . . . . . . . . . . . . . . . . 157

Lesson 24 . . . . . . . . . . . . . . . . . . . . . . . . . . . . . . . . . . . . . . . . . . . . . . . . . . . . . . . . . . . . . . . . . . 163

**Topic F: Addition and Subtraction of Fractions by Decomposition**

Lesson 25 . . . . . . . . . . . . . . . . . . . . . . . . . . . . . . . . . . . . . . . . . . . . . . . . . . . . . . . . . . . . . . . . . . 167

Lesson 26 . . . . . . . . . . . . . . . . . . . . . . . . . . . . . . . . . . . . . . . . . . . . . . . . . . . . . . . . . . . . . . . . . . 173

Lesson 27 . . . . . . . . . . . . . . . . . . . . . . . . . . . . . . . . . . . . . . . . . . . . . . . . . . . . . . . . . . . . . . . . . . 181

Lesson 28 . . . . . . . . . . . . . . . . . . . . . . . . . . . . . . . . . . . . . . . . . . . . . . . . . . . . . . . . . . . . . . . . . . 187

Lesson 29 . . . . . . . . . . . . . . . . . . . . . . . . . . . . . . . . . . . . . . . . . . . . . . . . . . . . . . . . . . . . . . . . . . 193

Lesson 30 . . . . . . . . . . . . . . . . . . . . . . . . . . . . . . . . . . . . . . . . . . . . . . . . . . . . . . . . . . . . . . . . . . 199

Lesson 31 . . . . . . . . . . . . . . . . . . . . . . . . . . . . . . . . . . . . . . . . . . . . . . . . . . . . . . . . . . . . . . . . . . 205

# A STORY OF UNITS – TEKS EDITION

## Lesson 1 Application Problem  4•5

Use your scissors to cut an index card on the diagonal lines. Prove that you have cut the rectangle into 4 fourths. Include a drawing in your explanation.

_____

_____

_____

_____

**Read**   **Draw**   **Write**

**Lesson 1:** Decompose fractions as a sum of unit fractions using strip diagrams.

Name _____  Date _____

1. Draw a number bond, and write the number sentence to match each strip diagram. The first one is done for you.

a.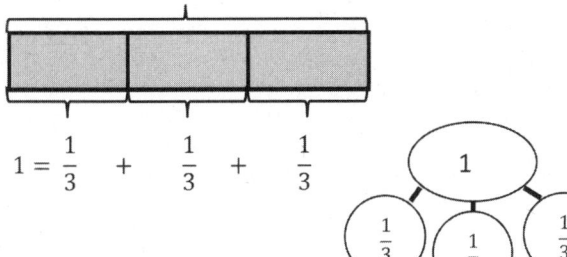

b.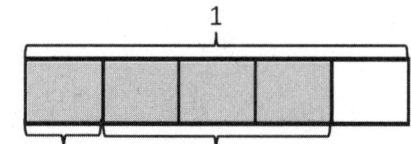

c.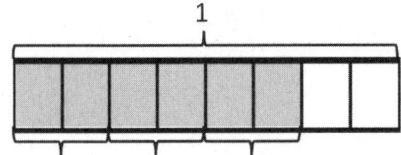

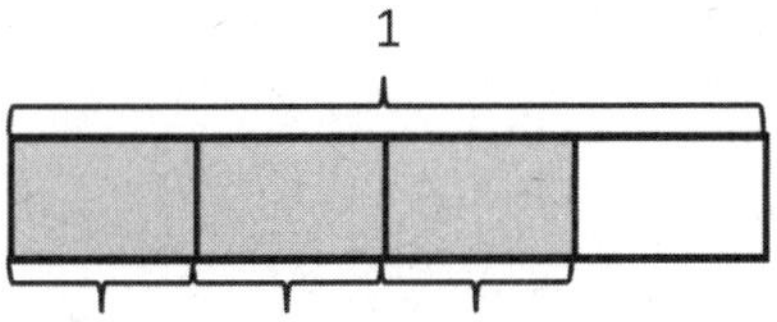

d.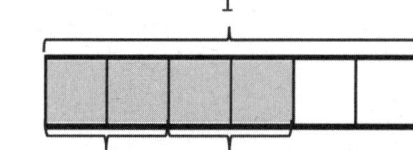

e.

f.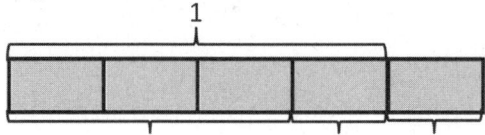

**Lesson 1:** Decompose fractions as a sum of unit fractions using strip diagrams.

g.

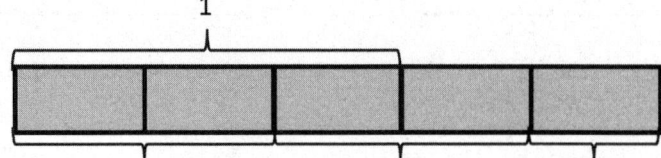

h.

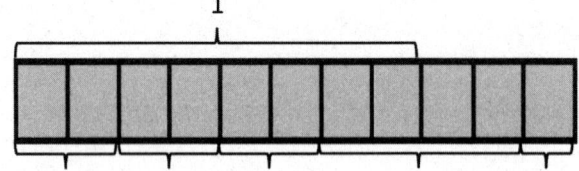

2. Draw and label strip diagrams to model each decomposition.

   a. $1 = \frac{1}{6} + \frac{1}{6} + \frac{1}{6} + \frac{1}{6} + \frac{1}{6} + \frac{1}{6}$

   b. $\frac{4}{5} = \frac{1}{5} + \frac{2}{5} + \frac{1}{5}$

   c. $\frac{7}{8} = \frac{3}{8} + \frac{3}{8} + \frac{1}{8}$

   d. $\frac{11}{8} = \frac{7}{8} + \frac{1}{8} + \frac{3}{8}$

A STORY OF UNITS – TEKS EDITION                                   Lesson 1 Problem Set  4•5

e. $\frac{12}{10} = \frac{6}{10} + \frac{4}{10} + \frac{2}{10}$

f. $\frac{15}{12} = \frac{8}{12} + \frac{3}{12} + \frac{4}{12}$

g. $1\frac{2}{3} = 1 + \frac{2}{3}$

h. $1\frac{5}{8} = 1 + \frac{1}{8} + \frac{1}{8} + \frac{3}{8}$

Lesson 1: Decompose fractions as a sum of unit fractions using strip diagrams.

# A STORY OF UNITS – TEKS EDITION

Lesson 1 Exit Ticket 4•5

Name _____ Date _____

1. Complete the number bond, and write the number sentence to match the strip diagram.

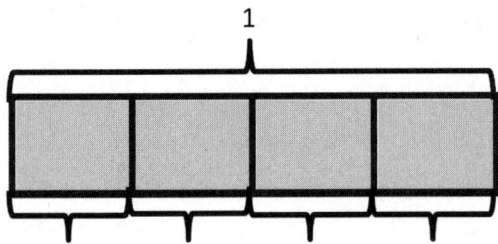
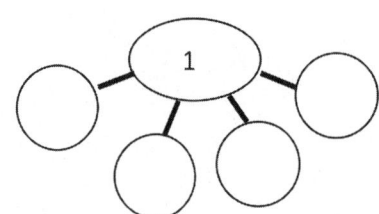

2. Draw and label strip diagrams to model each number sentence.

   a. $1 = \frac{1}{5} + \frac{1}{5} + \frac{1}{5} + \frac{1}{5} + \frac{1}{5}$

   b. $\frac{5}{6} = \frac{2}{6} + \frac{2}{6} + \frac{1}{6}$

Lesson 1: Decompose fractions as a sum of unit fractions using strip diagrams.

# A STORY OF UNITS – TEKS EDITION
## Lesson 2 Application Problem 4•5

Mrs. Salcido cut a small birthday cake into 6 equal pieces for 6 children. One child was not hungry, so she gave the birthday boy the extra piece. Draw a strip diagram to show how much cake each of the five children received.

_____

_____

_____

**Read**  **Draw**  **Write**

Lesson 2: Decompose fractions as a sum of unit fractions using strip diagrams.

# A STORY OF UNITS – TEKS EDITION

Lesson 2 Problem Set  4•5

Name _____  Date _____

1. Step 1: Draw and shade a strip diagram of the given fraction.
   Step 2: Record the decomposition as a sum of unit fractions.
   Step 3: Record the decomposition of the fraction two more ways.
   (The first one has been done for you.)

   a. $\frac{5}{8}$

   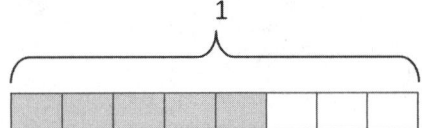

   $\frac{5}{8} = \frac{1}{8} + \frac{1}{8} + \frac{1}{8} + \frac{1}{8} + \frac{1}{8}$     $\frac{5}{8} = \frac{2}{8} + \frac{2}{8} + \frac{1}{8}$     $\frac{5}{8} = \frac{2}{8} + \frac{1}{8} + \frac{1}{8} + \frac{1}{8}$

   b. $\frac{9}{10}$

   c. $\frac{3}{2}$

Lesson 2: Decompose fractions as a sum of unit fractions using strip diagrams.

2. Step 1: Draw and shade a strip diagram of the given fraction.
   Step 2: Record the decomposition of the fraction in three different ways using number sentences.

   a. $\frac{7}{8}$

   b. $\frac{5}{3}$

   c. $\frac{7}{5}$

   d. $1\frac{1}{3}$

# A STORY OF UNITS – TEKS EDITION

Lesson 2 Exit Ticket 4•5

Name _____   Date _____

Step 1: Draw and shade a strip diagram of the given fraction.

Step 2: Record the decomposition of the fraction in three different ways using number sentences.

$\frac{4}{7}$

Lesson 2: Decompose fractions as a sum of unit fractions using strip diagrams.

# A STORY OF UNITS – TEKS EDITION

Lesson 3 Application Problem  4•5

A recipe calls for $\frac{3}{4}$ cup of milk. Saisha only has a $\frac{1}{4}$-cup measuring cup. If she doubles the recipe, how many times will she need to fill the $\frac{1}{4}$ cup with milk? Draw a strip diagram, and record as an addition sentence.

**Read**   **Draw**   **Write**

Lesson 3: Decompose fractions into sums of smaller unit fractions using strip diagrams.

Name _____ Date _____

1. The total length of each strip diagram represents 1. Decompose the shaded unit fractions as the sum of smaller unit fractions in at least two different ways. The first one has been done for you.

   a.

   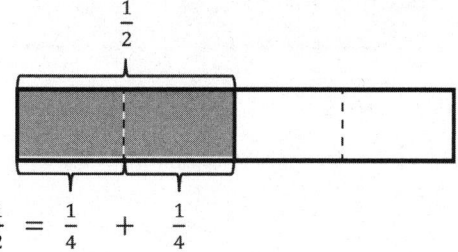

   $\frac{1}{2} = \frac{1}{4} + \frac{1}{4}$

   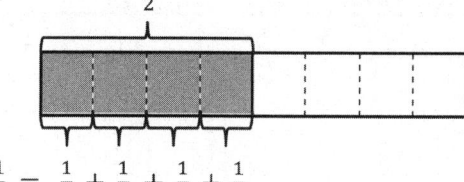

   $\frac{1}{2} = \frac{1}{8} + \frac{1}{8} + \frac{1}{8} + \frac{1}{8}$

   b.

   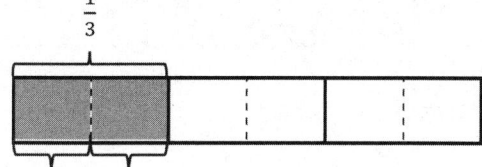

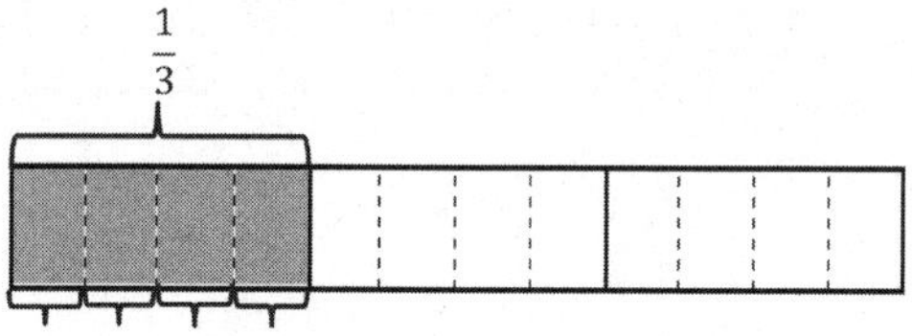

   c.

   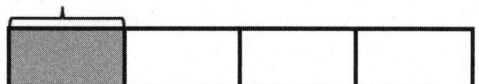

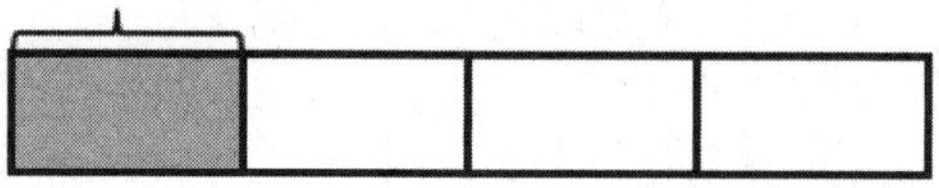

   d.

   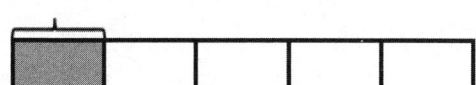   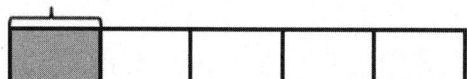

Lesson 3: Decompose fractions into sums of smaller unit fractions using strip diagrams.

2. The total length of each strip diagram represents 1. Decompose the shaded fractions as the sum of smaller unit fractions in at least two different ways.

   a.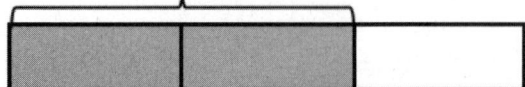

   b.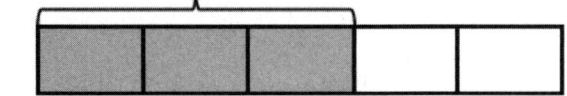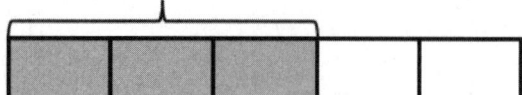

3. Draw and label strip diagrams to prove the following statements. The first one has been done for you.

   a. $\frac{2}{5} = \frac{4}{10}$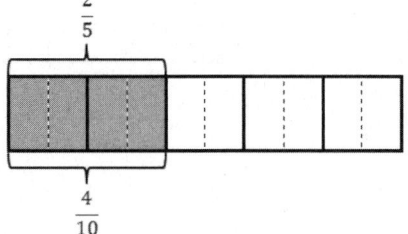

   b. $\frac{2}{6} = \frac{4}{12}$

Lesson 3: Decompose fractions into sums of smaller unit fractions using strip diagrams.

c. $\frac{3}{4} = \frac{6}{8}$

d. $\frac{3}{4} = \frac{9}{12}$

4. Show that $\frac{1}{2}$ is equivalent to $\frac{4}{8}$ using a strip diagram and a number sentence.

5. Show that $\frac{2}{3}$ is equivalent to $\frac{6}{9}$ using a strip diagram and a number sentence.

6. Show that $\frac{4}{6}$ is equivalent to $\frac{8}{12}$ using a strip diagram and a number sentence.

# A STORY OF UNITS – TEKS EDITION

Lesson 3 Exit Ticket  4•5

Name _____  Date _____

1. The total length of the strip diagram represents 1. Decompose the shaded unit fraction as the sum of smaller unit fractions in at least two different ways.

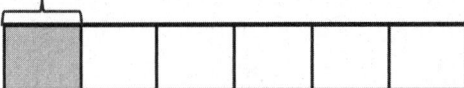   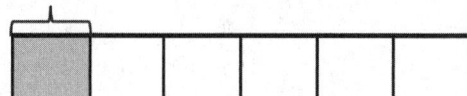

2. Draw a strip diagram to prove the following statement.

$$\frac{2}{3} = \frac{4}{6}$$

Lesson 3: Decompose fractions into sums of smaller unit fractions using strip diagrams.

21

# A STORY OF UNITS – TEKS EDITION

**Lesson 4 Application Problem** 4•5

A loaf of bread was cut into 6 equal slices. Each of the 6 slices was cut in half to make thinner slices for sandwiches. Mr. Beach used 4 slices. His daughter said, "Wow! You used $\frac{2}{6}$ of the loaf!" His son said, "No. He used $\frac{4}{12}$." Explain who was correct using a strip diagram.

**Read     Draw     Write**

Lesson 4: Decompose unit fractions using area models to show equivalence.

Name _____   Date _____

1. Draw horizontal lines to decompose each rectangle into the number of rows as indicated. Use the model to give the shaded area as a sum of unit fractions.

   a. 2 rows

   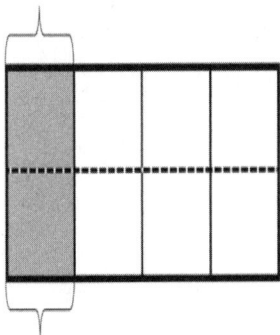

   $\dfrac{1}{4} = \dfrac{2}{\_\_}$

   $\dfrac{1}{4} = \dfrac{1}{8} + \dfrac{\_\_}{\_\_} = \dfrac{\_\_}{\_\_}$

   b. 2 rows

   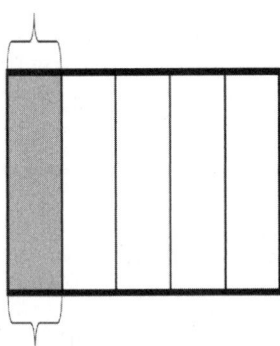

   c. 4 rows

   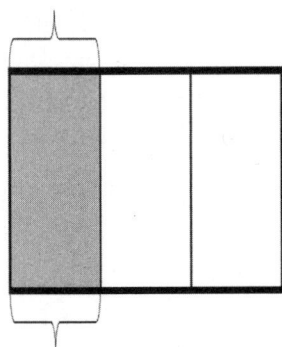

Lesson 4: Decompose unit fractions using area models to show equivalence.

2. Draw area models to show the decompositions represented by the number sentences below. Represent the decomposition as a sum of unit fractions.

   a. $\frac{1}{2} = \frac{3}{6}$

   b. $\frac{1}{2} = \frac{4}{8}$

   c. $\frac{1}{2} = \frac{5}{10}$

   d. $\frac{1}{3} = \frac{2}{6}$

   e. $\frac{1}{3} = \frac{4}{12}$

   f. $\frac{1}{4} = \frac{3}{12}$

3. Explain why $\frac{1}{12} + \frac{1}{12} + \frac{1}{12}$ is the same as $\frac{1}{4}$.

Lesson 4: Decompose unit fractions using area models to show equivalence.

A STORY OF UNITS – TEKS EDITION

Lesson 4 Exit Ticket  4•5

Name _____   Date _____

1. Draw horizontal lines to decompose each rectangle into the number of rows as indicated. Use the model to give the shaded area as a sum of unit fractions.

   a. 2 rows

   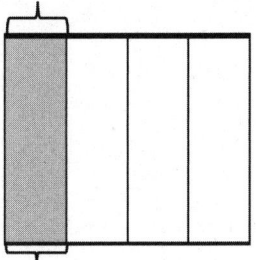

   b. 3 rows

   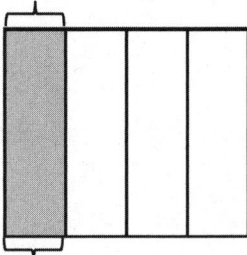

2. Draw an area model to show the decomposition represented by the number sentence below. Represent the decomposition as a sum of unit fractions.

   $\dfrac{3}{5} = \dfrac{6}{10}$

Lesson 4: Decompose unit fractions using area models to show equivalence.

Use area models to prove that $\frac{1}{2} = \frac{2}{4} = \frac{4}{8}$, $\frac{1}{2} = \frac{3}{6} = \frac{6}{12}$, and $\frac{1}{2} = \frac{5}{10}$. What conclusion can you make about $\frac{4}{8}$, $\frac{6}{12}$, and $\frac{5}{10}$? Explain.

_____

_____

_____

_____

**Read**  **Draw**  **Write**

Lesson 5: Decompose fractions using area models to show equivalence.

**Name** _____  **Date** _____

1. Each rectangle represents 1. Draw horizontal lines to decompose each rectangle into the fractional units as indicated. Use the model to give the shaded area as a sum of unit fractions. Use parentheses to show the relationship between the number sentences. The first one has been partially done for you.

   a. Sixths  $\frac{2}{3}$

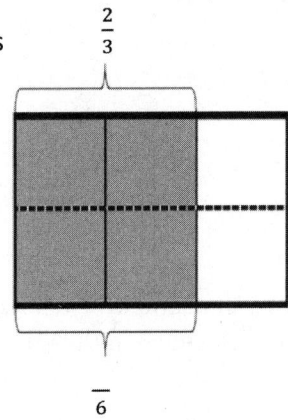

   $$\frac{2}{3} = \frac{4}{-}$$

   $$\frac{-}{3} + \frac{-}{3} = \left(\frac{1}{6} + \frac{1}{6}\right) + \left(\frac{1}{6} + \frac{1}{6}\right) = \frac{4}{-}$$

   $\frac{-}{6}$

   b. Tenths

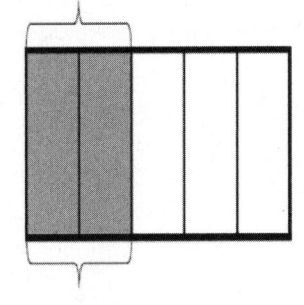

---

Lesson 5: Decompose fractions using area models to show equivalence.

c. Twelfths

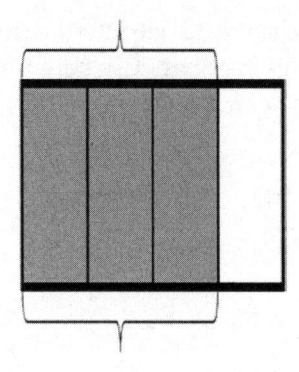

2. Draw area models to show the decompositions represented by the number sentences below. Express each as a sum of unit fractions. Use parentheses to show the relationship between the number sentences.

   a. $\frac{3}{5} = \frac{6}{10}$

   b. $\frac{3}{4} = \frac{6}{8}$

3. Step 1: Draw an area model for a fraction with units of thirds, fourths, or fifths.

   Step 2: Shade in more than one fractional unit.

   Step 3: Partition the area model again to find an equivalent fraction.

   Step 4: Write the equivalent fractions as a number sentence. (If you've written a number sentence like this one already on this Problem Set, start over.)

Name _____  Date _____

1. The rectangle below represents 1. Draw horizontal lines to decompose the rectangle into eighths. Use the model to give the shaded area as a sum of unit fractions. Use parentheses to show the relationship between the number sentences.

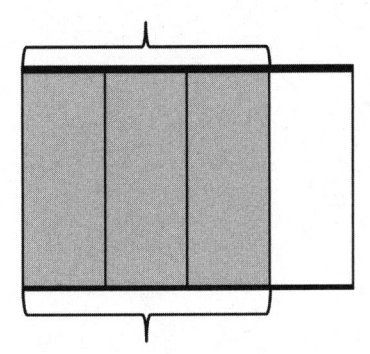

2. Draw an area model to show the decomposition represented by the number sentence below.

$$\frac{4}{5} = \frac{8}{10}$$

Lesson 5: Decompose fractions using area models to show equivalence.

Model an equivalent fraction for $\frac{4}{7}$ using an area model.

**Read**     **Draw**     **Write**

Lesson 6: Use the area model and multiplication to show the equivalence of two fractions.

# A STORY OF UNITS – TEKS EDITION

Lesson 6 Problem Set 4•5

Name _____ Date _____

Each rectangle represents 1.

1. The shaded unit fractions have been decomposed into smaller units. Express the equivalent fractions in a number sentence using multiplication. The first one has been done for you.

   a.

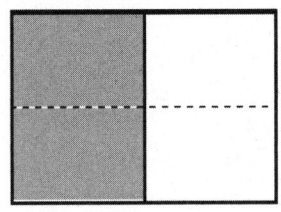

   $$\frac{1}{2} = \frac{1 \times 2}{2 \times 2} = \frac{2}{4}$$

   b.

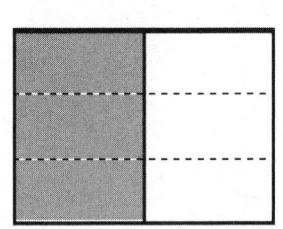

   c.

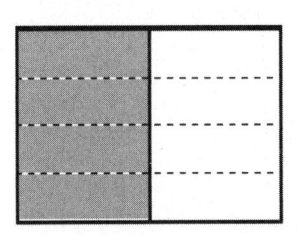

   d.

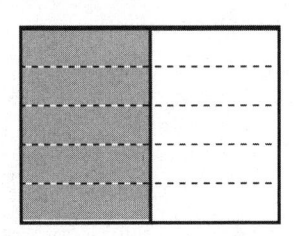

Lesson 6: Use the area model and multiplication to show the equivalence of two fractions.

2. Decompose the shaded fractions into smaller units using the area models. Express the equivalent fractions in a number sentence using multiplication.

   a.

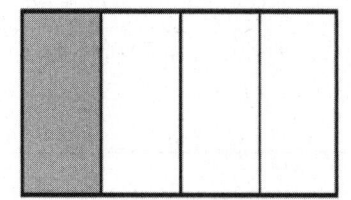

   b.

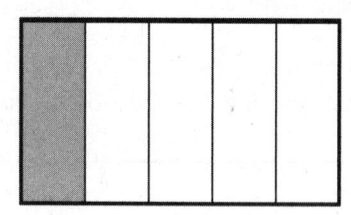

   c.

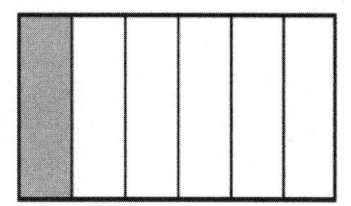

   d.

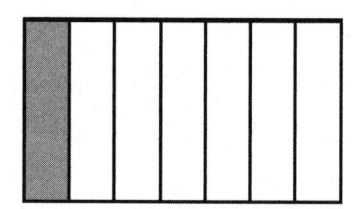

   e. What happened to the size of the fractional units when you decomposed the fraction?

   f. What happened to the total number of units in the whole when you decomposed the fraction?

3. Draw three different area models to represent 1 third by shading.
Decompose the shaded fraction into (a) sixths, (b) ninths, and (c) twelfths.
Use multiplication to show how each fraction is equivalent to 1 third.

   a.

   b.

   c.

**Name** _____ **Date** _____

Draw two different area models to represent 1 fourth by shading.
Decompose the shaded fraction into (a) eighths and (b) twelfths.
Use multiplication to show how each fraction is equivalent to 1 fourth.

a.

b.

Lesson 6: Use the area model and multiplication to show the equivalence of two fractions.

**A STORY OF UNITS – TEKS EDITION**  Lesson 7 Application Problem  4•5

Saisha gives some of her chocolate bar, pictured below, to her younger brother Lucas. He says, "Thanks for $\frac{3}{12}$ of the bar." Saisha responds, "No. I gave you $\frac{1}{4}$ of the bar." Explain why both Lucas and Saisha are correct.

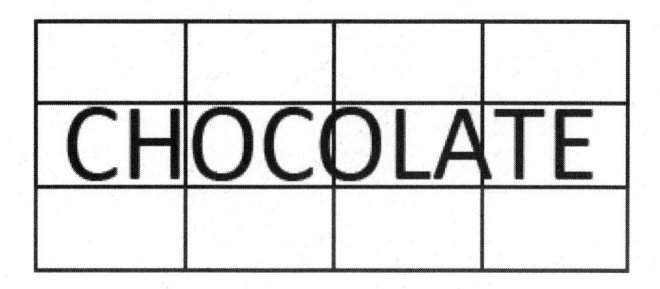

___

**Read**  **Draw**  **Write**

Lesson 7: Use the area model and multiplication to show the equivalence of two fractions.

A STORY OF UNITS – TEKS EDITION

Lesson 7 Problem Set 4•5

Name _____ Date _____

Each rectangle represents 1.

1. The shaded fractions have been decomposed into smaller units. Express the equivalent fractions in a number sentence using multiplication. The first one has been done for you.

   a.

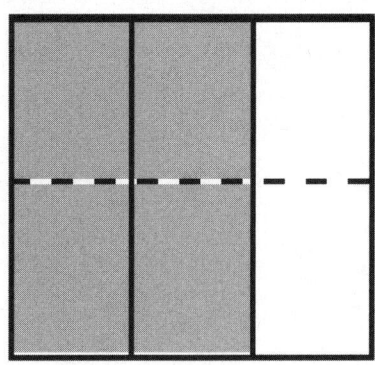

   $$\frac{2}{3} = \frac{2 \times 2}{3 \times 2} = \frac{4}{6}$$

   b.

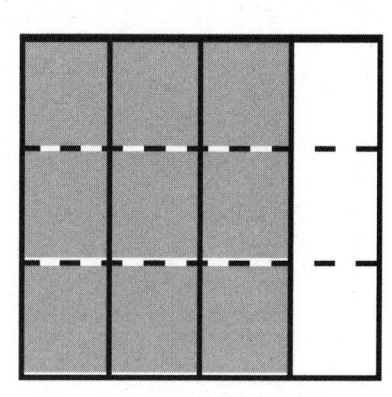

   c.

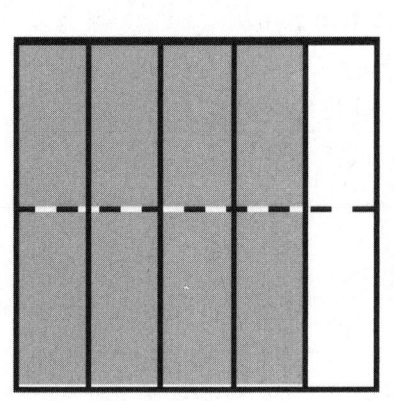

   d.

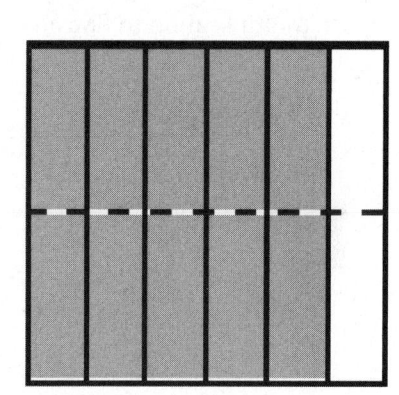

2. Decompose the shaded fractions into smaller units, as given below. Express the equivalent fractions in a number sentence using multiplication.

   a. Decompose into tenths.

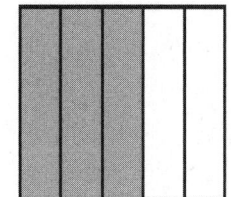

   b. Decompose into fifteenths.

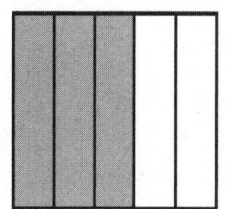

Lesson 7: Use the area model and multiplication to show the equivalence of two fractions.

47

3. Draw area models to prove that the following number sentences are true.

   a. $\frac{2}{5} = \frac{4}{10}$

   b. $\frac{2}{3} = \frac{8}{12}$

   c. $\frac{3}{6} = \frac{6}{12}$

   d. $\frac{4}{6} = \frac{8}{12}$

4. Use multiplication to find an equivalent fraction for each fraction below.

   a. $\frac{3}{4}$

   b. $\frac{4}{5}$

   c. $\frac{7}{6}$

   d. $\frac{12}{7}$

5. Determine which of the following are true number sentences. Correct those that are false by changing the right-hand side of the number sentence.

   a. $\frac{4}{3} = \frac{8}{9}$

   b. $\frac{5}{4} = \frac{10}{8}$

   c. $\frac{4}{5} = \frac{12}{10}$

   d. $\frac{4}{6} = \frac{12}{18}$

Lesson 7: Use the area model and multiplication to show the equivalence of two fractions.

Name _____ Date _____

1. Use multiplication to create an equivalent fraction for the fraction below.

$$\frac{2}{5}$$

2. Determine if the following is a true number sentence. If needed, correct the statement by changing the right-hand side of the number sentence.

$$\frac{3}{4} = \frac{9}{8}$$

What fraction of a foot is 1 inch? What fraction of a foot is 3 inches? (Hint: 12 inches = 1 foot.) Draw a strip diagram to model your work.

**Read**     **Draw**     **Write**

Lesson 8: Use the area model and division to show the equivalence of two fractions.

A STORY OF UNITS – TEKS EDITION

Lesson 8 Problem Set 4•5

Name _____  Date _____

Each rectangle represents 1.

1. Compose the shaded fractions into larger fractional units. Express the equivalent fractions in a number sentence using division. The first one has been done for you.

    a.

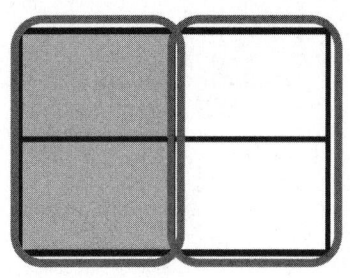

    $$\frac{2}{4} = \frac{2 \div 2}{4 \div 2} = \frac{1}{2}$$

    b.

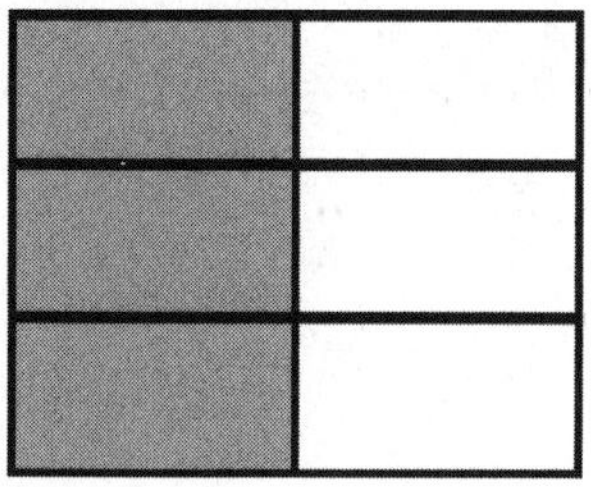

    c.

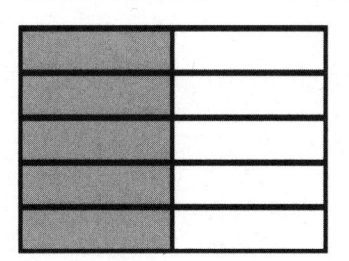

    d.

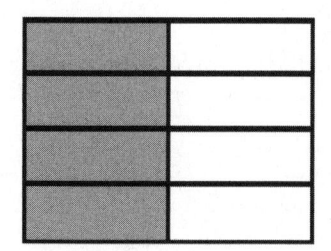

Lesson 8: Use the area model and division to show the equivalence of two fractions.

2. Compose the shaded fractions into larger fractional units. Express the equivalent fractions in a number sentence using division.

   a.

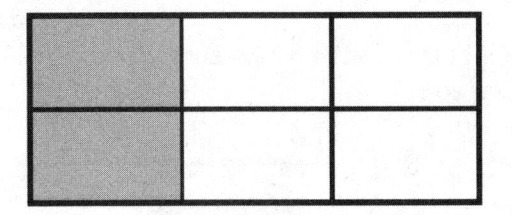

   b.

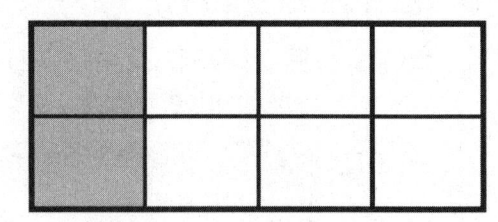

   c.

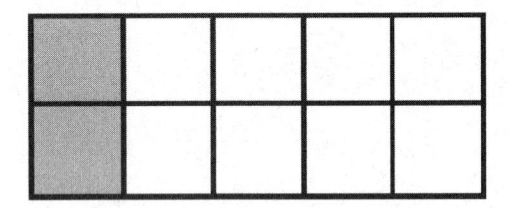

   d.

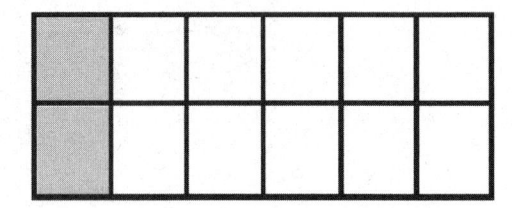

   e. What happened to the size of the fractional units when you composed the fraction?

   f. What happened to the total number of units in the whole when you composed the fraction?

3. a. In the first area model, show 2 sixths. In the second area model, show 3 ninths. Show how both fractions can be renamed as the same unit fraction.

b. Express the equivalent fractions in a number sentence using division.

4. a. In the first area model, show 2 eighths. In the second area model, show 3 twelfths. Show how both fractions can be composed, or renamed, as the same unit fraction.

b. Express the equivalent fractions in a number sentence using division.

Lesson 8: Use the area model and division to show the equivalence of two fractions.

Name _____  Date _____

a. In the first area model, show 2 sixths. In the second area model, show 4 twelfths. Show how both fractions can be composed, or renamed, as the same unit fraction.

b. Express the equivalent fractions in a number sentence using division.

**Nuri spent $\frac{9}{12}$ of his money on a book and the rest of his money on a pencil.**

a. Express how much of his money he spent on the pencil in fourths.

b. Nuri started with $1. How much did he spend on the pencil?

Read          Draw          Write

Lesson 9: Use the area model and division to show the equivalence of two fractions.

# A STORY OF UNITS – TEKS EDITION

Lesson 9 Problem Set 4•5

Name _____ Date _____

Each rectangle represents 1.

1. Compose the shaded fraction into larger fractional units. Express the equivalent fractions in a number sentence using division. The first one has been done for you.

a.

$$\frac{4}{6} = \frac{4 \div 2}{6 \div 2} = \frac{2}{3}$$

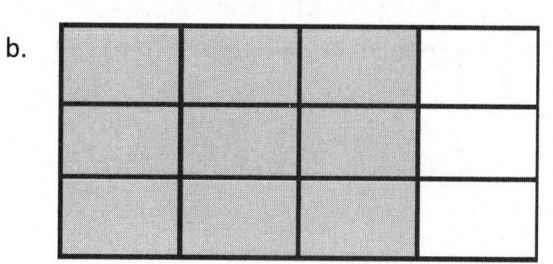

b.

c.

d.

**Lesson 9:** Use the area model and division to show the equivalence of two fractions.

2. Compose the shaded fractions into larger fractional units. Express the equivalent fractions in a number sentence using division.

   a.

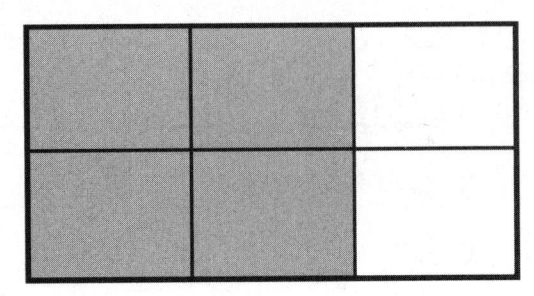

   b.

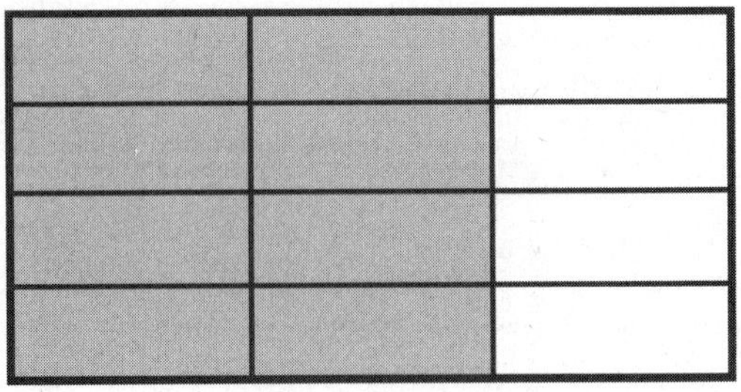

3. Draw an area model to represent each number sentence below.

   a. $\dfrac{4}{10} = \dfrac{4 \div 2}{10 \div 2} = \dfrac{2}{5}$

   b. $\dfrac{6}{9} = \dfrac{6 \div 3}{9 \div 3} = \dfrac{2}{3}$

Lesson 9: Use the area model and division to show the equivalence of two fractions.

4. Use division to rename each fraction given below. Draw a model if that helps you. See if you can use the largest common factor.

   a. $\frac{4}{8}$

   b. $\frac{12}{16}$

   c. $\frac{12}{20}$

   d. $\frac{16}{20}$

Name _____  Date _____

Draw an area model to show why the fractions are equivalent. Show the equivalence in a number sentence using division.

$$\frac{4}{10} = \frac{2}{5}$$

# Lesson 10 Application Problem

Kelly was baking bread but could only find her $\frac{1}{8}$-cup measuring cup. She needs $\frac{1}{4}$ cup sugar, $\frac{3}{4}$ cup whole wheat flour, and $\frac{1}{2}$ cup all-purpose flour. How many $\frac{1}{8}$ cups will she need for each ingredient?

___

___

___

___

**Read**　　　**Draw**　　　**Write**

**Lesson 10:** Explain fraction equivalence using a strip diagram and the number line, and relate that to the use of multiplication and division.

Name _____  Date _____

1. Label each number line with the fractions shown on the strip diagram. Circle the fraction that labels the point on the number line that also names the shaded part of the strip diagram.

   a.

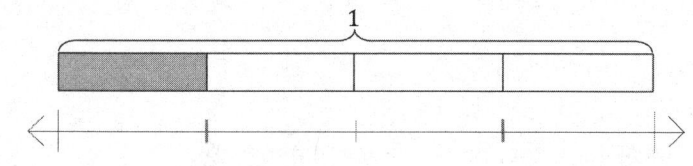

   b.

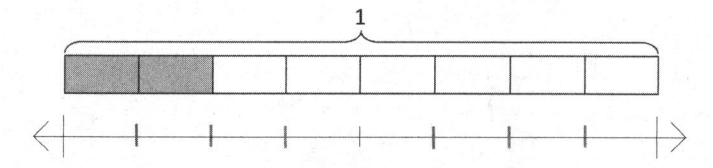

   c.

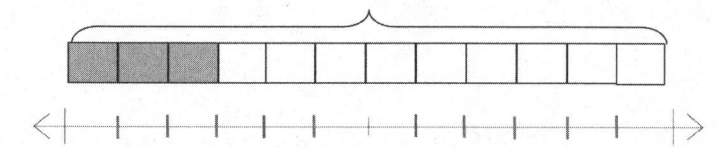

**Lesson 10:** Explain fraction equivalence using a strip diagram and the number line, and relate that to the use of multiplication and division.

2. Write number sentences using multiplication to show:

   a. The fraction represented in 1(a) is equivalent to the fraction represented in 1(b).

   b. The fraction represented in 1(a) is equivalent to the fraction represented in 1(c).

3. Use each shaded strip diagram below as a ruler to draw a number line. Mark each number line with the fractional units shown on the strip diagram, and circle the fraction that labels the point on the number line that also names the shaded part of the strip diagram.

   a.

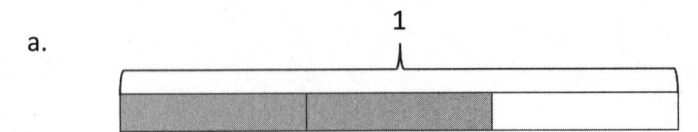

   b.

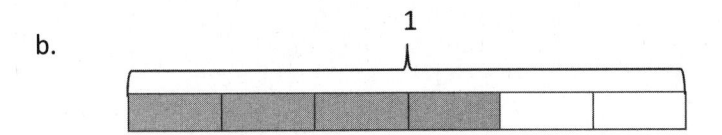

   c.

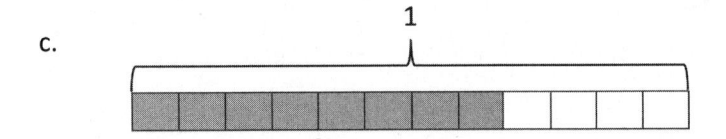

Lesson 10: Explain fraction equivalence using a strip diagram and the number line, and relate that to the use of multiplication and division.

4. Write number sentences using division to show:

   a. The fraction represented in 3(a) is equivalent to the fraction represented in 3(b).

   b. The fraction represented in 3(a) is equivalent to the fraction represented in 3(c).

5. a. Partition a number line from 0 to 1 into fifths. Decompose $\frac{2}{5}$ into 4 equal lengths.

   b. Write a number sentence using multiplication to show what fraction represented on the number line is equivalent to $\frac{2}{5}$.

   c. Write a number sentence using division to show what fraction represented on the number line is equivalent to $\frac{2}{5}$.

Lesson 10: Explain fraction equivalence using a strip diagram and the number line, and relate that to the use of multiplication and division.

**A STORY OF UNITS – TEKS EDITION**  Lesson 10 Exit Ticket 4•5

Name _____  Date _____

1. Partition a number line from 0 to 1 into sixths. Decompose $\frac{2}{6}$ into 4 equal lengths.

2. Write a number sentence using multiplication to show what fraction represented on the number line is equivalent to $\frac{2}{6}$.

3. Write a number sentence using division to show what fraction represented on the number line is equivalent to $\frac{2}{6}$.

**Lesson 10:** Explain fraction equivalence using a strip diagram and the number line, and relate that to the use of multiplication and division.

Plot $\frac{1}{4}$, $\frac{4}{5}$, and $\frac{5}{8}$ on a number line, and compare the three points.

**Read**  **Draw**  **Write**

Lesson 11: Reason using benchmarks to compare two fractions on the number line.

Name _____    Date _____

1. a. Plot the following points on the number line without measuring.

   i. $\frac{1}{3}$    ii. $\frac{5}{6}$    iii. $\frac{7}{12}$

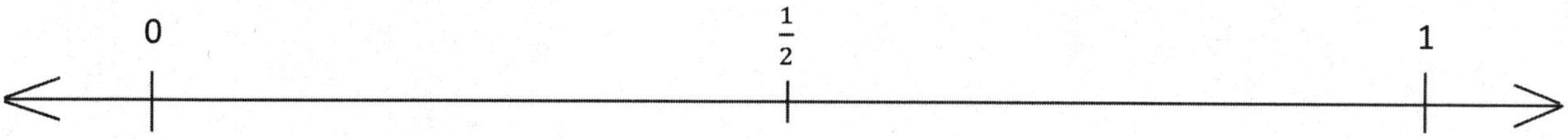

   ←———|————————————|————————————|———→
       0            $\frac{1}{2}$           1

   b. Use the number line in Part (a) to compare the fractions by writing >, <, or = on the lines.

   i. $\frac{7}{12}$ _____ $\frac{1}{2}$    ii. $\frac{7}{12}$ _____ $\frac{5}{6}$

2. a. Plot the following points on the number line without measuring.

   i. $\frac{11}{12}$    ii. $\frac{1}{4}$    iii. $\frac{3}{8}$

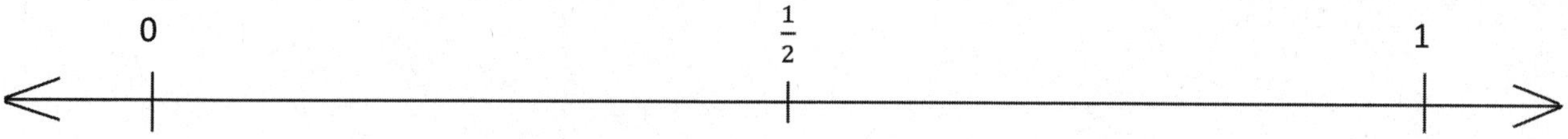

   ←———|————————————|————————————|———→
       0            $\frac{1}{2}$           1

   b. Select two fractions from Part (a), and use the given number line to compare them by writing >, <, or =.

   c. Explain how you plotted the points in Part (a).

Lesson 11: Reason using benchmarks to compare two fractions on the number line.

3. Compare the fractions given below by writing > or < on the lines.
   Give a brief explanation for each answer referring to the benchmarks 0, $\frac{1}{2}$, and 1.

   a. $\frac{1}{2}$ _____ $\frac{3}{4}$

   b. $\frac{1}{2}$ _____ $\frac{7}{8}$

   c. $\frac{2}{3}$ _____ $\frac{2}{5}$

   d. $\frac{9}{10}$ _____ $\frac{3}{5}$

   e. $\frac{2}{3}$ _____ $\frac{7}{8}$

   f. $\frac{1}{3}$ _____ $\frac{2}{4}$

   g. $\frac{2}{3}$ _____ $\frac{5}{10}$

   h. $\frac{11}{12}$ _____ $\frac{2}{5}$

   i. $\frac{49}{100}$ _____ $\frac{51}{100}$

   j. $\frac{7}{16}$ _____ $\frac{51}{100}$

Name _____  Date _____

1. Plot the following points on the number line without measuring.

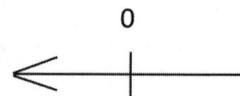

   a. $\frac{8}{10}$         b. $\frac{3}{5}$         c. $\frac{1}{4}$

   ←――――|―――――――――――|―――――――――――|――――→
        0            $\frac{1}{2}$            1

2. Use the number line in Problem 1 to compare the fractions by writing >, <, or = on the lines.

   a. $\frac{1}{4}$ _____ $\frac{1}{2}$

   b. $\frac{8}{10}$ _____ $\frac{3}{5}$

   c. $\frac{1}{2}$ _____ $\frac{3}{5}$

   d. $\frac{1}{4}$ _____ $\frac{8}{10}$

# A STORY OF UNITS – TEKS EDITION

Lesson 11 Template  4•5

*Application Problem*

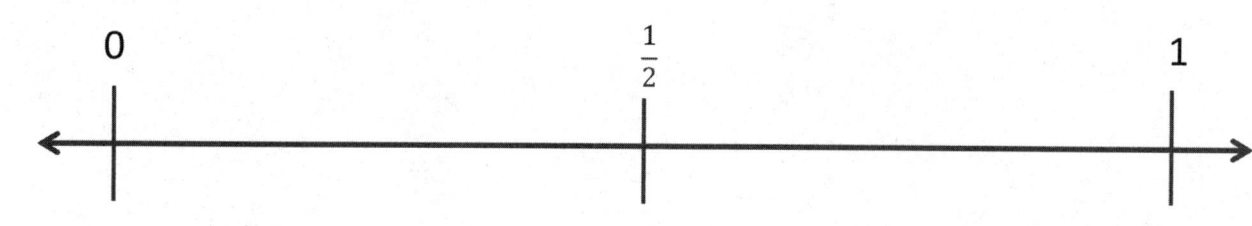

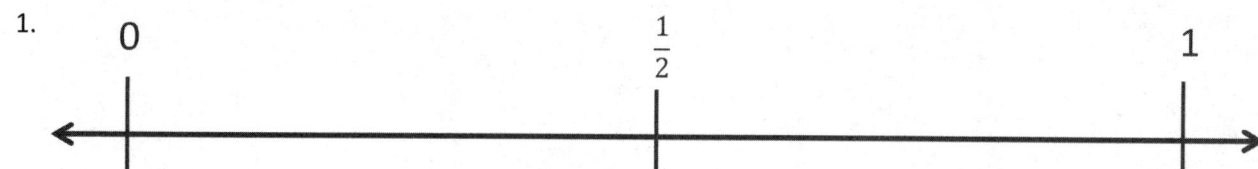

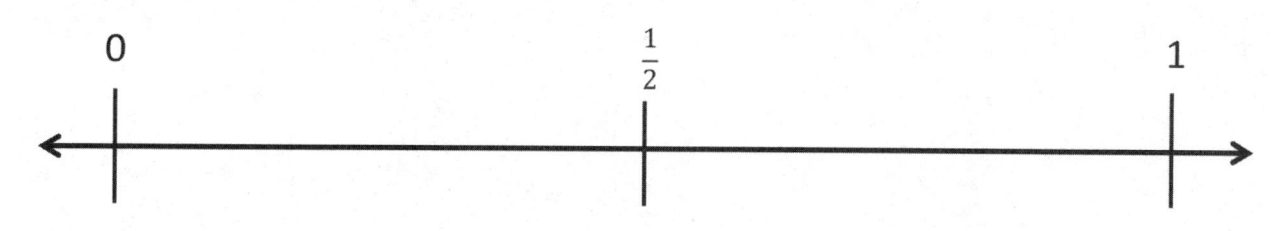

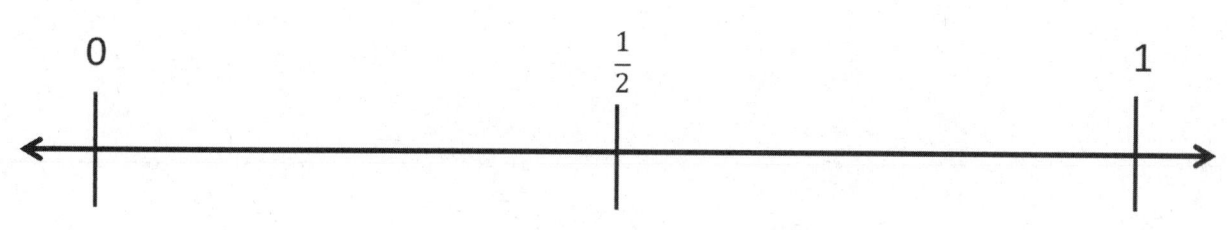

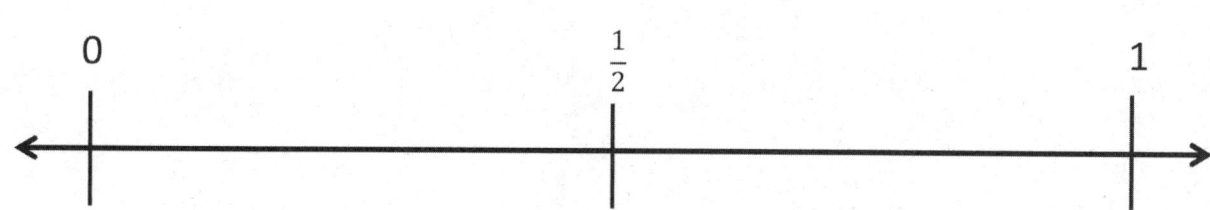

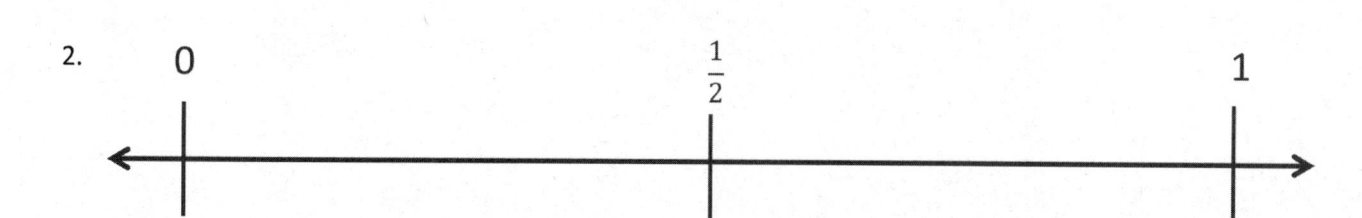

```
0              1/2              1
|---------------|---------------|
```

1.
```
0              1/2              1
|---------------|---------------|
```

```
0              1/2              1
|---------------|---------------|
```

```
0              1/2              1
|---------------|---------------|
```

```
0              1/2              1
|---------------|---------------|
```

2.
```
0              1/2              1
|---------------|---------------|
```

number line

**Lesson 11:** Reason using benchmarks to compare two fractions on the number line.

81

Mr. and Mrs. Reynolds went for a run. Mr. Reynolds ran for $\frac{6}{10}$ mile. Mrs. Reynolds ran for $\frac{2}{5}$ mile. Who ran farther? Explain how you know. Use the benchmarks 0, $\frac{1}{2}$, and 1 to explain your answer.

**Read**      **Draw**      **Write**

Lesson 12: Reason using benchmarks to compare two fractions on the number line.

A STORY OF UNITS – TEKS EDITION

Lesson 12 Problem Set 4•5

Name _____  Date _____

1. Place the following fractions on the number line given.

   a. $\frac{4}{3}$    b. $\frac{11}{6}$    c. $\frac{17}{12}$

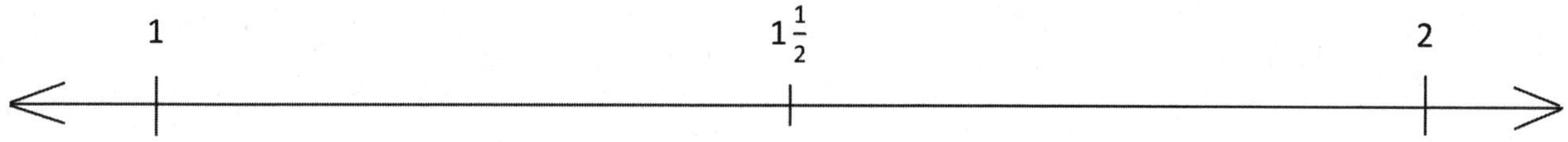

   <--|----------------|----------------|--->
      1              $1\frac{1}{2}$       2

2. Use the number line in Problem 1 to compare the fractions by writing >, <, or = on the lines.

   a. $1\frac{5}{6}$ _____ $1\frac{5}{12}$    b. $1\frac{1}{3}$ _____ $1\frac{5}{12}$

3. Place the following fractions on the number line given.

   a. $\frac{11}{8}$    b. $\frac{7}{4}$    c. $\frac{15}{12}$

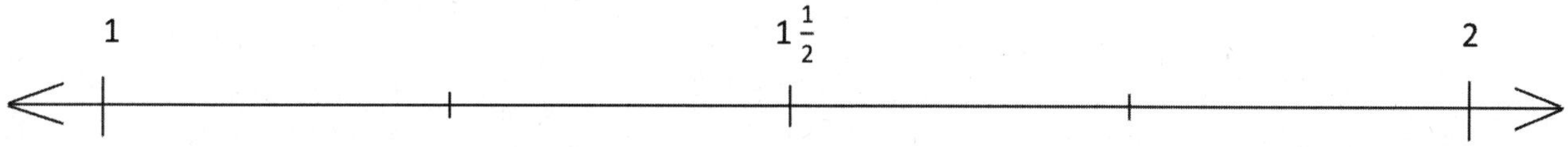

   <--|--------|--------|--------|--------|-->
      1               $1\frac{1}{2}$              2

4. Use the number line in Problem 3 to explain the reasoning you used when determining whether $\frac{11}{8}$ or $\frac{15}{12}$ is greater.

Lesson 12: Reason using benchmarks to compare two fractions on the number line.

85

# A STORY OF UNITS – TEKS EDITION

**Lesson 12 Problem Set** 4•5

5. Compare the fractions given below by writing > or < on the lines. Give a brief explanation for each answer referring to benchmarks.

a. $\dfrac{3}{8}$ _____ $\dfrac{7}{12}$

b. $\dfrac{5}{12}$ _____ $\dfrac{7}{8}$

c. $\dfrac{8}{6}$ _____ $\dfrac{11}{12}$

d. $\dfrac{5}{12}$ _____ $\dfrac{1}{3}$

e. $\dfrac{7}{5}$ _____ $\dfrac{11}{10}$

f. $\dfrac{5}{4}$ _____ $\dfrac{7}{8}$

g. $\dfrac{13}{12}$ _____ $\dfrac{9}{10}$

h. $\dfrac{6}{8}$ _____ $\dfrac{5}{4}$

i. $\dfrac{8}{12}$ _____ $\dfrac{8}{4}$

j. $\dfrac{7}{5}$ _____ $\dfrac{16}{10}$

---

Lesson 12: Reason using benchmarks to compare two fractions on the number line.

**A STORY OF UNITS – TEKS EDITION**  Lesson 12 Exit Ticket  4•5

Name _____  Date _____

1. Place the following fractions on the number line given.

   a. $\frac{5}{4}$   b. $\frac{10}{7}$   c. $\frac{16}{9}$

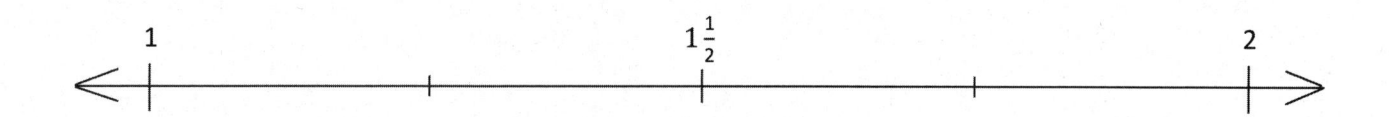

   ```
   <---|-------|-------|-------|-------|--->
       1              1½              2
   ```

2. Compare the fractions using >, <, or =.

   a. $\frac{5}{4}$ _____ $\frac{10}{7}$   b. $\frac{5}{4}$ _____ $\frac{16}{9}$   c. $\frac{16}{9}$ _____ $\frac{10}{7}$

Lesson 12: Reason using benchmarks to compare two fractions on the number line.

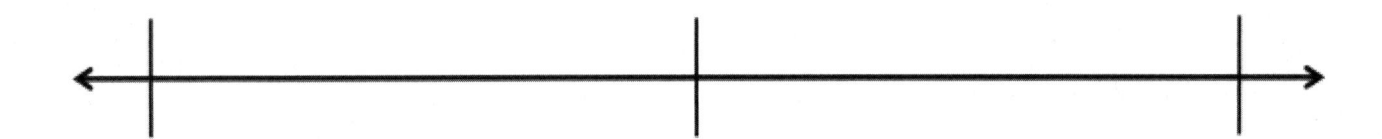

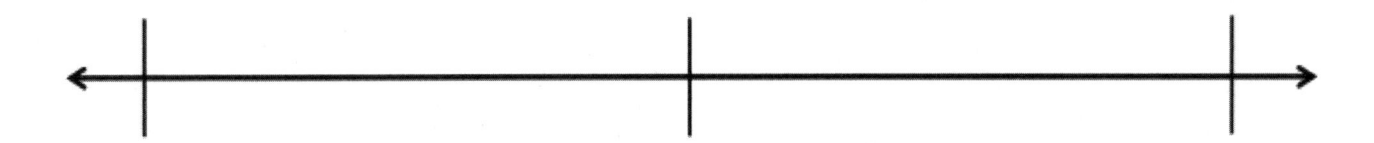

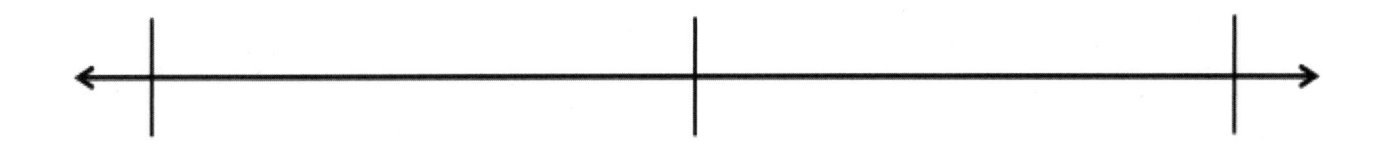

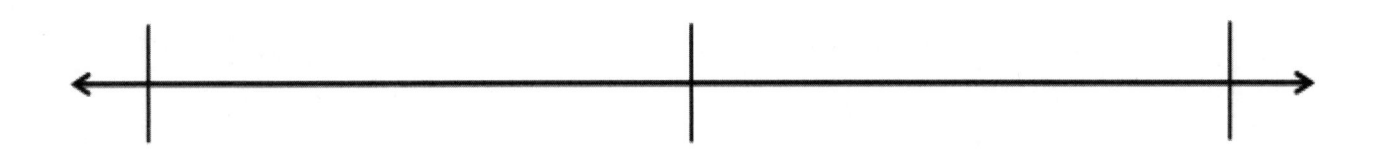

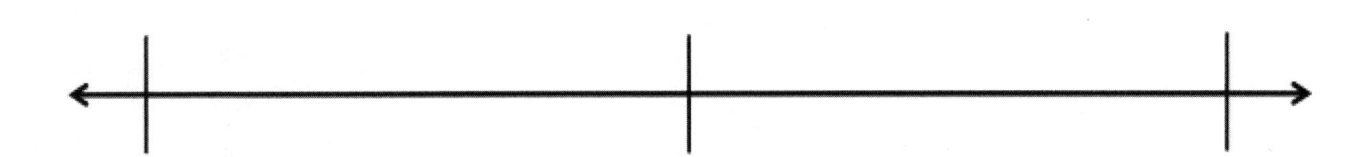

blank number lines with midpoint

Lesson 12: Reason using benchmarks to compare two fractions on the number line.

Compare $\frac{4}{5}$, $\frac{3}{4}$, and $\frac{9}{10}$ using <, >, or =. Explain your reasoning using a benchmark.

**Read**  **Draw**  **Write**

Lesson 13: Find common units or number of units to compare two fractions.

Name _____   Date _____

1. Compare the pairs of fractions by reasoning about the size of the units. Use >, <, or =.

   a.  1 fourth _____ 1 fifth

   b.  3 fourths _____ 3 fifths

   c.  1 tenth _____ 1 twelfth

   d.  7 tenths _____ 7 twelfths

2. Compare by reasoning about the following pairs of fractions with the same or related numerators. Use >, <, or =. Explain your thinking using words, pictures, or numbers. Problem 2(b) has been done for you.

   a.  $\frac{3}{5}$ _____ $\frac{3}{4}$

   b.  $\frac{2}{5} < \frac{4}{9}$

   because $\frac{2}{5} = \frac{4}{10}$

   4 tenths is less than 4 ninths because tenths are smaller than ninths.

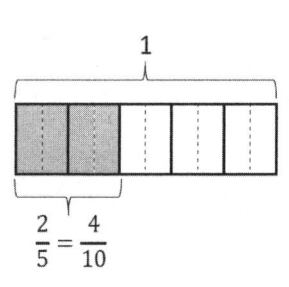

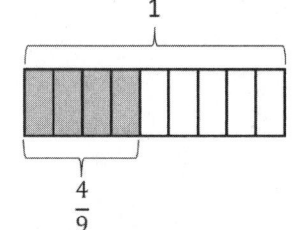

   c.  $\frac{7}{11}$ _____ $\frac{7}{13}$

   d.  $\frac{6}{7}$ _____ $\frac{12}{15}$

Lesson 13: Find common units or number of units to compare two fractions.

3. Draw two strip diagrams to model each pair of the following fractions with related denominators. Use >, <, or = to compare.

   a. $\frac{2}{3}$ _____ $\frac{5}{6}$

   b. $\frac{3}{4}$ _____ $\frac{7}{8}$

   c. $1\frac{3}{4}$ _____ $1\frac{7}{12}$

A STORY OF UNITS – TEKS EDITION                    Lesson 13 Problem Set   4•5

4. Draw one number line to model each pair of fractions with related denominators. Use >, <, or = to compare.

   a. $\frac{2}{3}$ _____ $\frac{5}{6}$

   b. $\frac{3}{8}$ _____ $\frac{1}{4}$

   c. $\frac{2}{6}$ _____ $\frac{5}{12}$

   d. $\frac{8}{9}$ _____ $\frac{2}{3}$

5. Compare each pair of fractions using >, <, or =. Draw a model if you choose to.

   a. $\frac{3}{4}$ _____ $\frac{3}{7}$

   b. $\frac{4}{5}$ _____ $\frac{8}{12}$

   c. $\frac{7}{10}$ _____ $\frac{3}{5}$

   d. $\frac{2}{3}$ _____ $\frac{11}{15}$

   e. $\frac{3}{4}$ _____ $\frac{11}{12}$

   f. $\frac{7}{3}$ _____ $\frac{7}{4}$

   g. $1\frac{1}{3}$ _____ $1\frac{2}{9}$

   h. $1\frac{2}{3}$ _____ $1\frac{4}{7}$

Lesson 13:  Find common units or number of units to compare two fractions.

95

6. Timmy drew the picture to the right and claimed that $\frac{2}{3}$ is less than $\frac{7}{12}$. Evan says he thinks $\frac{2}{3}$ is greater than $\frac{7}{12}$. Who is correct? Support your answer with a picture.

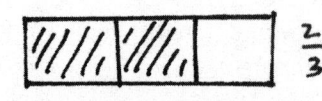

# A STORY OF UNITS – TEKS EDITION

Lesson 13 Exit Ticket 4•5

Name _____  Date _____

1. Draw strip diagrams to compare the following fractions:

   $\frac{2}{5}$ _____ $\frac{3}{10}$

2. Use a number line to compare the following fractions:

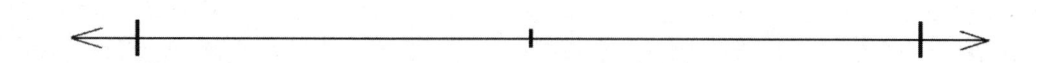

   $\frac{4}{3}$ _____ $\frac{7}{6}$

Lesson 13: Find common units or number of units to compare two fractions.

Jamal ran $\frac{2}{3}$ mile. Ming ran $\frac{2}{4}$ mile. Laina ran $\frac{7}{12}$ mile. Who ran the farthest? What do you think is the easiest way to determine the answer to this question?

_____

_____

_____

**Read**     **Draw**     **Write**

**Lesson 14:** Find common units or number of units to compare two fractions.

A STORY OF UNITS – TEKS EDITION

Lesson 14 Problem Set 4•5

Name _____ Date _____

1. Draw an area model for each pair of fractions, and use it to compare the two fractions by writing >, <, or = on the line. The first two have been partially done for you. Each rectangle represents 1.

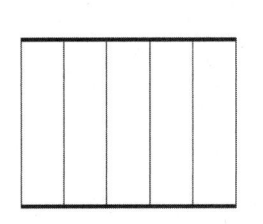

a. $\frac{1}{2}$ __<__ $\frac{2}{3}$

$\frac{1 \times 3}{2 \times 3} = \frac{3}{6}$

$\frac{2 \times 2}{3 \times 2} = \frac{4}{6}$

b. $\frac{4}{5}$ _____ $\frac{3}{4}$

c. $\frac{3}{5}$ _____ $\frac{4}{7}$

d. $\frac{3}{7}$ _____ $\frac{2}{6}$

e. $\frac{5}{8}$ _____ $\frac{6}{9}$

f. $\frac{2}{3}$ _____ $\frac{3}{4}$

Lesson 14: Find common units or number of units to compare two fractions.

101

2. Rename the fractions, as needed, using multiplication in order to compare each pair of fractions by writing >, <, or =.

   a. $\dfrac{3}{5}$ _____ $\dfrac{5}{6}$

   b. $\dfrac{2}{6}$ _____ $\dfrac{3}{8}$

   c. $\dfrac{7}{5}$ _____ $\dfrac{10}{8}$

   d. $\dfrac{4}{3}$ _____ $\dfrac{6}{5}$

3. Use any method to compare the fractions. Record your answer using >, <, or =.

   a. $\dfrac{3}{4}$ _____ $\dfrac{7}{8}$

   b. $\dfrac{6}{8}$ _____ $\dfrac{3}{5}$

   c. $\dfrac{6}{4}$ _____ $\dfrac{8}{6}$

   d. $\dfrac{8}{5}$ _____ $\dfrac{9}{6}$

4. Explain two ways you have learned to compare fractions. Provide evidence using words, pictures, or numbers.

A STORY OF UNITS – TEKS EDITION　　　　　　　　　　　Lesson 14 Exit Ticket　4•5

Name _____　　Date _____

Draw an area model for each pair of fractions, and use it to compare the two fractions by writing >, <, or = on the line.

1. $\frac{3}{4}$ _____ $\frac{4}{5}$

2. $\frac{2}{6}$ _____ $\frac{3}{5}$

Lesson 14:　Find common units or number of units to compare two fractions.

Keisha ran $\frac{5}{6}$ mile in the morning and $\frac{2}{3}$ mile in the afternoon. Did Keisha run farther in the morning or in the afternoon? Explain.

___

___

___

___

**Read**      **Draw**      **Write**

**Lesson 15:** Use visual models to add and subtract two fractions with the same units.

Name _____ Date _____

1. Solve.

   a. 3 fifths − 1 fifth = _____
   
   b. 5 fifths − 3 fifths = _____
   
   c. 3 halves − 2 halves = _____
   
   d. 6 fourths − 3 fourths = _____

2. Solve.

   a. $\frac{5}{6} - \frac{3}{6}$
   
   b. $\frac{6}{8} - \frac{4}{8}$
   
   c. $\frac{3}{10} - \frac{3}{10}$
   
   d. $\frac{5}{5} - \frac{4}{5}$
   
   e. $\frac{5}{4} - \frac{4}{4}$
   
   f. $\frac{5}{4} - \frac{3}{4}$

3. Solve. Use a number bond to show how to convert the difference to a mixed number. Problem (a) has been completed for you.

   a. $\frac{12}{8} - \frac{3}{8} = \frac{9}{8} = 1\frac{1}{8}$
   
   (number bond: $\frac{9}{8}$ → $\frac{8}{8}$ and $\frac{1}{8}$)
   
   b. $\frac{12}{6} - \frac{5}{6}$
   
   c. $\frac{9}{5} - \frac{3}{5}$
   
   d. $\frac{14}{8} - \frac{3}{8}$
   
   e. $\frac{8}{4} - \frac{2}{4}$
   
   f. $\frac{15}{10} - \frac{3}{10}$

Lesson 15: Use visual models to add and subtract two fractions with the same units.

4. Solve. Write the sum in unit form.

   a. 2 fourths + 1 fourth = _____

   b. 4 fifths + 3 fifths = _____

5. Solve.

   a. $\frac{2}{8} + \frac{5}{8}$

   b. $\frac{4}{12} + \frac{5}{12}$

6. Solve. Use a number bond to decompose the sum. Record your final answer as a mixed number. Problem (a) has been completed for you.

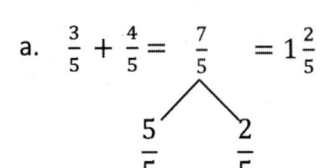

   a. $\frac{3}{5} + \frac{4}{5} = \frac{7}{5} = 1\frac{2}{5}$

   (number bond: $\frac{7}{5}$ decomposed into $\frac{5}{5}$ and $\frac{2}{5}$)

   b. $\frac{4}{4} + \frac{3}{4}$

   c. $\frac{6}{9} + \frac{6}{9}$

   d. $\frac{7}{10} + \frac{6}{10}$

   e. $\frac{5}{6} + \frac{7}{6}$

   f. $\frac{9}{8} + \frac{5}{8}$

7. Solve. Use a number line to model your answer.

   a. $\frac{7}{4} - \frac{5}{4}$

   b. $\frac{5}{4} + \frac{2}{4}$

Name _____ Date _____

1. Solve. Use a number bond to decompose the difference. Record your final answer as a mixed number.

   $\frac{16}{9} - \frac{5}{9}$

2. Solve. Use a number bond to decompose the sum. Record your final answer as a mixed number.

   $\frac{5}{12} + \frac{10}{12}$

Name _____  Date _____

blank number lines

**Lesson 15:** Use visual models to add and subtract two fractions with the same units.

Use a number bond to show the relationship between $\frac{2}{3}$, $\frac{3}{6}$, and $\frac{5}{6}$. Then, use the fractions to write two addition and two subtraction sentences.

**Read**     **Draw**     **Write**

Lesson 16: Use visual models to add and subtract two fractions with the same units, including subtracting from one whole.

Name _____ Date _____

1. Use the following three fractions to write two subtraction and two addition number sentences.

   a.  $\frac{8}{5}, \frac{2}{5}, \frac{10}{5}$

   b.  $\frac{15}{8}, \frac{7}{8}, \frac{8}{8}$

2. Solve. Model each subtraction problem with a number line, and solve by both counting up and subtracting. Part (a) has been completed for you.

   a.  $1 - \frac{3}{4}$

   $\frac{4}{4} - \frac{3}{4} = \frac{1}{4}$

   b.  $1 - \frac{8}{10}$

   c.  $1 - \frac{3}{5}$

   d.  $1 - \frac{5}{8}$

   e.  $1\frac{2}{10} - \frac{7}{10}$

   f.  $1\frac{1}{5} - \frac{3}{5}$

Lesson 16: Use visual models to add and subtract two fractions with the same units, including subtracting from one whole.

3. Find the difference in two ways. Use number bonds to decompose the total. Part (a) has been completed for you.

a. $1\frac{2}{5} - \frac{4}{5}$

Number bond: $1\frac{2}{5}$ decomposes into $\frac{5}{5}$ and $\frac{2}{5}$

$\frac{5}{5} + \frac{2}{5} = \frac{7}{5}$

$\frac{7}{5} - \frac{4}{5} = \boxed{\frac{3}{5}}$

$\frac{5}{5} - \frac{4}{5} = \frac{1}{5}$

$\frac{1}{5} + \frac{2}{5} = \boxed{\frac{3}{5}}$

b. $1\frac{3}{6} - \frac{4}{6}$

c. $1\frac{6}{8} - \frac{7}{8}$

d. $1\frac{1}{10} - \frac{7}{10}$

e. $1\frac{3}{12} - \frac{6}{12}$

Name _____  Date _____

1. Solve. Model the problem with a number line, and solve by both counting up and subtracting.

   $1 - \frac{2}{5}$

2. Find the difference in two ways. Use a number bond to show the decomposition.

   $1\frac{2}{7} - \frac{5}{7}$

# A STORY OF UNITS – TEKS EDITION

Lesson 17 Practice Sheet  4•5

Name _____  Date _____

| Problem A: | $\frac{1}{8} + \frac{3}{8} + \frac{4}{8}$ |
|---|---|

| Problem B: | $\frac{1}{6} + \frac{4}{6} + \frac{2}{6}$ |
|---|---|

| Problem C: | $\frac{11}{10} - \frac{4}{10} - \frac{1}{10}$ |
|---|---|

adding and subtracting fractions

**Lesson 17:** Add and subtract more than two fractions.

121

Problem D: $1 - \frac{3}{12} - \frac{5}{12}$

Problem E: $\frac{5}{8} + \frac{4}{8} + \frac{1}{8}$

Problem F: $1\frac{1}{5} - \frac{2}{5} - \frac{3}{5}$

adding and subtracting fractions

**Lesson 17:** Add and subtract more than two fractions.

# A STORY OF UNITS – TEKS EDITION

Lesson 17 Problem Set 4•5

Name _____ Date _____

1. Show one way to solve each problem. Express sums and differences as a mixed number when possible. Use number bonds when it helps you. Part (a) is partially completed.

| a. $\frac{2}{5} + \frac{3}{5} + \frac{1}{5}$ $= \frac{5}{5} + \frac{1}{5} = 1 + \frac{1}{5}$ $= \underline{\hspace{1cm}}$ | b. $\frac{3}{6} + \frac{1}{6} + \frac{3}{6}$ | c. $\frac{5}{7} + \frac{7}{7} + \frac{2}{7}$ |
|---|---|---|
| d. $\frac{7}{8} - \frac{3}{8} - \frac{1}{8}$ | e. $\frac{7}{9} + \frac{1}{9} + \frac{4}{9}$ | f. $\frac{4}{10} + \frac{11}{10} + \frac{5}{10}$ |
| g. $1 - \frac{3}{12} - \frac{4}{12}$ | h. $1\frac{2}{3} - \frac{1}{3} - \frac{1}{3}$ | i. $\frac{10}{12} + \frac{5}{12} + \frac{2}{12} + \frac{7}{12}$ |

Lesson 17: Add and subtract more than two fractions.

2. Monica and Stuart used different strategies to solve $\frac{5}{8} + \frac{2}{8} + \frac{5}{8}$.

**Monica's Way**

$$\frac{5}{8} + \frac{2}{8} + \frac{5}{8} = \frac{7}{8} + \frac{5}{8} = \frac{8}{8} + \frac{4}{8} = 1\frac{4}{8}$$

$\frac{1}{8}$   $\frac{4}{8}$

**Stuart's Way**

$$\frac{5}{8} + \frac{2}{8} + \frac{5}{8} = \frac{12}{8} = 1 + \frac{4}{8} = 1\frac{4}{8}$$

$\frac{8}{8}$   $\frac{4}{8}$

Whose strategy do you like best? Why?

3. You gave one solution for each part of Problem 1. Now, for each problem indicated below, give a different solution method.

1(c)     $\frac{5}{7} + \frac{7}{7} + \frac{2}{7}$

1(f)     $\frac{4}{10} + \frac{11}{10} + \frac{5}{10}$

1(g)     $1 - \frac{3}{12} - \frac{4}{12}$

Name _____  Date _____

Solve the following problems. Use number bonds to help you.

1. $\frac{5}{9} + \frac{2}{9} + \frac{4}{9}$

2. $1 - \frac{5}{8} - \frac{1}{8}$

Fractions are all around us! Make a list of times that you have used fractions, heard fractions, or seen fractions. Be ready to share your ideas.

_____

_____

_____

_____

**Read**      **Draw**      **Write**

Lesson 18: Solve word problems involving addition and subtraction of fractions.

# A STORY OF UNITS – TEKS EDITION

Lesson 18 Problem Set  4•5

Name _____  Date _____

Use the RDW process to solve.

1. Sue ran $\frac{9}{10}$ mile on Monday and $\frac{7}{10}$ mile on Tuesday. How many miles did Sue run in the 2 days?

2. Mr. Salazar cut his son's birthday cake into 8 equal pieces. Mr. Salazar, Mrs. Salazar, and the birthday boy each ate 1 piece of cake. What fraction of the cake was left?

3. Maria spent $\frac{4}{7}$ of her money on a book and saved the rest. What fraction of her money did Maria save?

Lesson 18: Solve word problems involving addition and subtraction of fractions.

4. Mrs. Jones had $1\frac{4}{8}$ pizzas left after a party. After giving some to Gary, she had $\frac{7}{8}$ pizza left. What fraction of a pizza did she give Gary?

5. A baker had 2 pans of corn bread. He served $1\frac{1}{4}$ pans. What fraction of a pan was left?

6. Marius combined $\frac{4}{8}$ gallon of lemonade, $\frac{3}{8}$ gallon of cranberry juice, and $\frac{6}{8}$ gallon of soda water to make punch for a party. How many gallons of punch did he make in all?

Name _____    Date _____

Use the RDW process to solve.

1. Mrs. Smith took her bird to the vet. Tweety weighed $1\frac{3}{10}$ pounds. The vet said that Tweety weighed $\frac{4}{10}$ pound more last year. How much did Tweety weigh last year?

2. Hudson picked $1\frac{1}{4}$ baskets of apples. Suzy picked 2 baskets of apples. How many more baskets of apples did Suzy pick than Hudson?

Winnie went shopping and spent $\frac{2}{5}$ of the money that was on a gift card. What fraction of the money was left on the card? Draw a number line and a number bond to help show your thinking.

_____

_____

_____

**Read** **Draw** **Write**

Lesson 19: Add a fraction less than 1 to, or subtract a fraction less than 1 from, a whole number using decomposition and visual models.

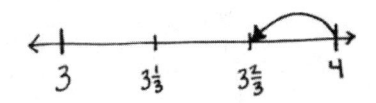

A STORY OF UNITS – TEKS EDITION                           Lesson 19 Problem Set  4•5

Name _____    Date _____

1. Draw a strip diagram to match each number sentence. Then, complete the number sentence.

   a.  $3 + \frac{1}{3} =$ _____      b.  $4 + \frac{3}{4} =$ _____

   c.  $3 - \frac{1}{4} =$ _____      d.  $5 - \frac{2}{5} =$ _____

2. Use the following three numbers to write two subtraction and two addition number sentences.

   a.  $6, 6\frac{3}{8}, \frac{3}{8}$      b.  $\frac{4}{7}, 9, 8\frac{3}{7}$

3. Solve using a number bond. Draw a number line to represent each number sentence. The first one has been done for you.

   a.  $4 - \frac{1}{3} = 3\frac{2}{3}$      b.  $5 - \frac{2}{3} =$ _____

   $4 - \frac{1}{3} = 3\frac{2}{3}$
   
   $3 \quad \frac{3}{3}$

   (number line from 3 to 4 showing $3\frac{1}{3}$, $3\frac{2}{3}$)

Lesson 19: Add a fraction less than 1 to, or subtract a fraction less than 1 from, a whole number using decomposition and visual models.

c. $7 - \frac{3}{8} =$ _____

d. $10 - \frac{4}{10} =$ _____

4. Complete the subtraction sentences using number bonds.

a. $3 - \frac{1}{10} =$ _____

b. $5 - \frac{3}{4} =$ _____

c. $6 - \frac{5}{8} =$ _____

d. $7 - \frac{3}{9} =$ _____

e. $8 - \frac{6}{10} =$ _____

f. $29 - \frac{9}{12} =$ _____

Name _____ Date _____

Complete the subtraction sentences using number bonds. Draw a model if needed.

1. $6 - \frac{1}{5} =$ _____

2. $8 - \frac{5}{6} =$ _____

3. $7 - \frac{5}{8} =$ _____

Shelly read her book for $\frac{1}{2}$ hour each afternoon for 9 days. How many hours did Shelly spend reading in all 9 days?

**Read****Draw****Write**

Name _____   Date _____

1. Rename each fraction as a mixed number by decomposing it into two parts as shown below. Model the decomposition with a number line and a number bond.

   a. $\frac{11}{3}$

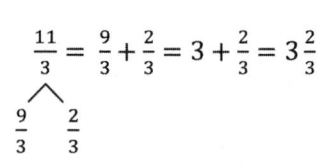

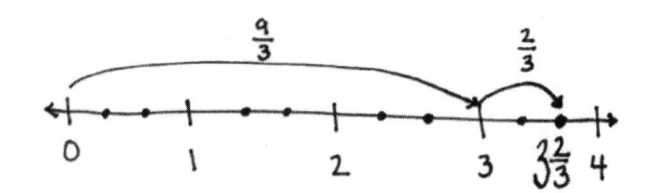

   $\frac{11}{3} = \frac{9}{3} + \frac{2}{3} = 3 + \frac{2}{3} = 3\frac{2}{3}$

   b. $\frac{12}{5}$

   c. $\frac{13}{2}$

   d. $\frac{15}{4}$

2. Convert each fraction to a mixed number.

| a. $\frac{9}{4} =$ | b. $\frac{17}{5} =$ | c. $\frac{25}{6} =$ |
|---|---|---|
| d. $\frac{30}{7} =$ | e. $\frac{38}{8} =$ | f. $\frac{48}{9} =$ |
| g. $\frac{63}{10} =$ | h. $\frac{84}{10} =$ | i. $\frac{37}{12} =$ |

Name _____ Date _____

1. Rename the fraction as a mixed number by decomposing it into two parts. Model the decomposition with a number line and a number bond.

$$\frac{17}{5}$$

2. Convert the fraction to a mixed number. Model with a number line.

$\frac{19}{3}$

3. Convert the fraction to a mixed number.

$\frac{11}{4}$

Lesson 20: Decompose and compose fractions greater than 1 to express them in various forms.

**Lesson 21 Application Problem**

Mrs. Fowler knew that the perimeter of the soccer field was $\frac{1}{6}$ mile. Her goal was to walk two miles while watching her daughter's game. If she walked around the field 13 times, did she meet her goal? Explain your thinking.

_____

_____

_____

**Read**    **Draw**    **Write**

Lesson 21: Decompose and compose fractions greater than 1 to express them in various forms.

Name _____  Date _____

1. Convert each mixed number to a fraction greater than 1. Draw a number line to model your work.

   a. $3\frac{1}{4}$

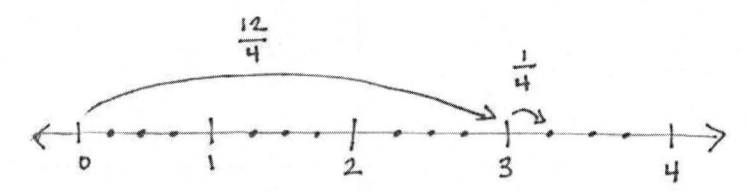

   [Number line from 0 to 4 with tick marks at quarters; arrow from 0 spanning $\frac{12}{4}$ to 3, then $\frac{1}{4}$ more]

   $3\frac{1}{4} = 3 + \frac{1}{4} = \frac{12}{4} + \frac{1}{4} = \frac{13}{4}$

   b. $2\frac{4}{5}$

   c. $3\frac{5}{8}$

   d. $4\frac{4}{10}$

   e. $4\frac{7}{9}$

Lesson 21: Decompose and compose fractions greater than 1 to express them in various forms.

147

2. Convert each mixed number to a fraction greater than 1.

| a. $2\frac{3}{4}$ | b. $2\frac{2}{5}$ | c. $3\frac{3}{6}$ |
|---|---|---|
| d. $3\frac{3}{8}$ | e. $3\frac{1}{10}$ | f. $4\frac{3}{8}$ |
| g. $5\frac{2}{3}$ | h. $6\frac{1}{2}$ | i. $7\frac{3}{10}$ |

Name _____   Date _____

Convert each mixed number to a fraction greater than 1.

1. $3\frac{1}{5}$

2. $2\frac{3}{5}$

3. $4\frac{2}{9}$

Barbara needed $3\frac{1}{4}$ cups of flour for her recipe. If she measured $\frac{1}{4}$ cup at a time, how many times did she have to fill the measuring cup?

**Read**     **Draw**     **Write**

Lesson 22: Compare fractions greater than 1 by reasoning using benchmark fractions.

Name _____ Date _____

1. a. Plot the following points on the number line without measuring.

   i. $2\frac{7}{8}$   ii. $3\frac{1}{6}$   iii. $\frac{25}{12}$

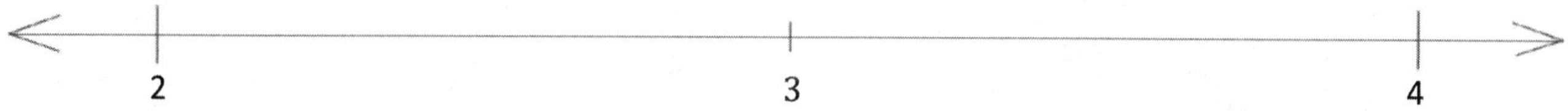

   <---+-------------------+-------------------+---->
       2                   3                   4

   b. Use the number line in Problem 1(a) to compare the fractions by writing >, <, or =.

   i. $\frac{29}{12}$ _____ $2\frac{7}{8}$   ii. $\frac{29}{12}$ _____ $3\frac{1}{6}$

2. a. Plot the following points on the number line without measuring.

   i. $\frac{70}{9}$   ii. $8\frac{2}{4}$   iii. $\frac{25}{3}$

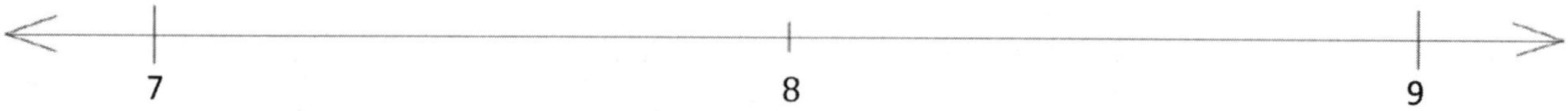

   <---+-------------------+-------------------+---->
       7                   8                   9

   b. Compare the following by writing >, <, or =.

   i. $8\frac{2}{4}$ _____ $\frac{25}{3}$   ii. $\frac{70}{9}$ _____ $8\frac{2}{4}$

   c. Explain how you plotted the points in Problem 2(a).

Lesson 22: Compare fractions greater than 1 by reasoning using benchmark fractions.

3. Compare the fractions given below by writing >, <, or =. Give a brief explanation for each answer, referring to benchmark fractions.

a. $5\frac{1}{3}$ _____ $4\frac{3}{4}$

b. $\frac{12}{6}$ _____ $\frac{25}{12}$

c. $\frac{18}{7}$ _____ $\frac{17}{5}$

d. $5\frac{2}{5}$ _____ $5\frac{5}{8}$

e. $6\frac{2}{3}$ _____ $6\frac{3}{7}$

f. $\frac{31}{7}$ _____ $\frac{32}{8}$

g. $\frac{31}{10}$ _____ $\frac{25}{8}$

h. $\frac{39}{12}$ _____ $\frac{19}{6}$

i. $\frac{49}{50}$ _____ $3\frac{90}{100}$

j. $5\frac{5}{12}$ _____ $5\frac{51}{100}$

Name _____ Date _____

Compare the fractions given below by writing >, <, or =.

Give a brief explanation for each answer, referring to benchmark fractions.

1.  $3\frac{2}{3}$ _____ $3\frac{4}{6}$

2.  $\frac{12}{3}$ _____ $\frac{27}{7}$

3.  $\frac{10}{6}$ _____ $\frac{5}{4}$

4.  $3\frac{2}{5}$ _____ $3\frac{3}{10}$

Lesson 22: Compare fractions greater than 1 by reasoning using benchmark fractions.

Jeremy ran 27 laps on a track that was $\frac{1}{8}$ mile long.  Jimmy ran 15 laps on a track that was $\frac{1}{4}$ mile long.  Who ran farther?

**Read**  **Draw**  **Write**

**Lesson 23:** Compare fractions greater than 1 by creating common numerators or denominators.

Name _____  Date _____

1. Draw a strip diagram to model each comparison. Use >, <, or = to compare.

   a.  $3\frac{2}{3}$ _____ $3\frac{5}{6}$

   b.  $3\frac{2}{5}$ _____ $3\frac{6}{10}$

   c.  $4\frac{3}{6}$ _____ $4\frac{1}{3}$

   d.  $4\frac{5}{8}$ _____ $\frac{19}{4}$

2. Use an area model to make like units. Then, use >, <, or = to compare.

   a.  $2\frac{3}{5}$ _____ $\frac{18}{7}$

   b.  $2\frac{3}{8}$ _____ $2\frac{1}{3}$

**Lesson 23:** Compare fractions greater than 1 by creating common numerators or denominators.

3. Compare each pair of fractions using >, <, or = using any strategy.

   a. $5\frac{3}{4}$ _____ $5\frac{3}{8}$

   b. $5\frac{2}{5}$ _____ $5\frac{8}{10}$

   c. $5\frac{6}{10}$ _____ $\frac{27}{5}$

   d. $5\frac{2}{3}$ _____ $5\frac{9}{15}$

   e. $\frac{7}{2}$ _____ $\frac{7}{3}$

   f. $\frac{12}{3}$ _____ $\frac{15}{4}$

   g. $\frac{22}{5}$ _____ $4\frac{2}{7}$

   h. $\frac{21}{4}$ _____ $5\frac{2}{5}$

   i. $\frac{29}{8}$ _____ $\frac{11}{3}$

   j. $3\frac{3}{4}$ _____ $3\frac{4}{7}$

# A STORY OF UNITS – TEKS EDITION

Lesson 23 Exit Ticket 4•5

Name _____  Date _____

Compare each pair of fractions using >, <, or = using any strategy.

1. $4\frac{3}{8}$ _____ $4\frac{1}{4}$

2. $3\frac{4}{5}$ _____ $3\frac{9}{10}$

3. $2\frac{1}{3}$ _____ $2\frac{2}{5}$

4. $10\frac{2}{5}$ _____ $10\frac{3}{4}$

**Lesson 23:** Compare fractions greater than 1 by creating common numerators or denominators.

Name _____  Date _____

1. The chart to the right shows the distance fourth graders in Ms. Smith's class were able to run before stopping for a rest. Create a dot plot to display the data in the table.

| Student | Distance (in miles) |
|---------|---------------------|
| Joe | $2\frac{1}{2}$ |
| Arianna | $1\frac{3}{4}$ |
| Bobbi | $2\frac{1}{8}$ |
| Morgan | $1\frac{5}{8}$ |
| Jack | $2\frac{5}{8}$ |
| Saisha | $2\frac{1}{4}$ |
| Tyler | $2\frac{2}{4}$ |
| Jenny | $\frac{5}{8}$ |
| Anson | $2\frac{2}{8}$ |
| Chandra | $2\frac{4}{8}$ |

Lesson 24: Solve word problems with dot plots.

2. Solve each problem.

   a. Who ran a mile farther than Jenny?

   b. Who ran a mile less than Jack?

   c. Two students ran exactly $2\frac{1}{4}$ miles. Identify the students. How many quarter miles did each student run?

   d. What is the difference, in miles, between the longest and shortest distance run?

   e. Compare the distances run by Arianna and Morgan using >, <, or =.

   f. Ms. Smith ran twice as far as Jenny. How far did Ms. Smith run? Write her distance as a mixed number.

   g. Mr. Reynolds ran $1\frac{3}{10}$ miles. Use >, <, or = to compare the distance Mr. Reynolds ran to the distance that Ms. Smith ran. Who ran farther?

3. Using the information in the table and on the dot plot, develop and write a question similar to those above. Solve, and then ask your partner to solve. Did you solve in the same way? Did you get the same answer?

Name _____ Date _____

Mr. O'Neil asked his students to record the length of time they read over the weekend. The times are listed in the table.

1. At the bottom of the page, make a dot plot of the data.

2. One of the students read $\frac{3}{4}$ hour on Friday, $\frac{3}{4}$ hour on Saturday, and $\frac{3}{4}$ hour on Sunday. How many hours did that student read over the weekend? Name that student.

| Student | Length of time (in hours) |
|---|---|
| Robin | $\frac{1}{2}$ |
| Bill | 1 |
| Katrina | $\frac{3}{4}$ |
| Kelly | $1\frac{3}{4}$ |
| Mary | $1\frac{1}{2}$ |
| Gail | $2\frac{1}{4}$ |
| Scott | $1\frac{3}{4}$ |
| Ben | $2\frac{2}{4}$ |

Lesson 24: Solve word problems with dot plots.

Both Allison and Jennifer jogged on Sunday. When asked about their distances, Allison said, "I ran $2\frac{7}{8}$ miles this morning and $3\frac{3}{8}$ miles this afternoon. So, I ran a total of about 6 miles," and Jennifer said, "I ran $3\frac{1}{10}$ miles this morning and $3\frac{3}{10}$ miles this evening. I ran a total of $6\frac{4}{10}$ miles." How do their answers differ?

Read     Draw     Write

Lesson 25: Estimate sums and differences using benchmark numbers.

Name _____ Date _____

1. Estimate each sum or difference to the nearest half or whole number by rounding. Explain your estimate using words or a number line.

   a. $2\frac{1}{12} + 1\frac{7}{8} \approx$ _____

   b. $1\frac{11}{12} + 5\frac{3}{4} \approx$ _____

   c. $8\frac{7}{8} - 2\frac{1}{9} \approx$ _____

   d. $6\frac{1}{8} - 2\frac{1}{12} \approx$ _____

   e. $3\frac{3}{8} + 5\frac{1}{9} \approx$ _____

Lesson 25: Estimate sums and differences using benchmark numbers.

# A STORY OF UNITS – TEKS EDITION

Lesson 25 Problem Set 4•5

2. Estimate each sum or difference to the nearest half or whole number by rounding. Explain your estimate using words or a number line.

   a. $\frac{16}{5} + \frac{11}{4} \approx$ _____

   b. $\frac{17}{3} - \frac{15}{7} \approx$ _____

   c. $\frac{59}{10} + \frac{26}{10} \approx$ _____

3. Montoya's estimate for $8\frac{5}{8} - 2\frac{1}{3}$ was 7. Julio's estimate was $6\frac{1}{2}$. Whose estimate do you think is closer to the actual difference? Explain.

4. Use benchmark numbers or mental math to estimate the sum or difference.

| a. $14\frac{3}{4} + 29\frac{11}{12}$ | b. $3\frac{5}{12} + 54\frac{5}{8}$ |
|---|---|
| c. $17\frac{4}{5} - 8\frac{7}{12}$ | d. $\frac{65}{8} - \frac{37}{6}$ |

Lesson 25: Estimate sums and differences using benchmark numbers.

Name _____  Date _____

Estimate each sum or difference to the nearest half or whole number by rounding. Explain your estimate using words or a number line.

1. $2\frac{9}{10} + 2\frac{1}{4} \approx$ _____

2. $11\frac{8}{9} - 3\frac{3}{8} \approx$ _____

One board measures 2 meters 70 centimeters. Another measures 87 centimeters. What is the total length of the two boards expressed in meters and centimeters?

**Read**     **Draw**     **Write**

Lesson 26: Add a mixed number and a fraction.

# A STORY OF UNITS – TEKS EDITION

Lesson 26 Problem Set  4•5

Name _____  Date _____

1. Solve.

   a.  $3\frac{1}{4} + \frac{1}{4}$

   b.  $7\frac{3}{4} + \frac{1}{4}$

   c.  $\frac{3}{8} + 5\frac{2}{8}$

   d.  $\frac{1}{8} + 6\frac{7}{8}$

2. Complete the number sentences.

| | | | |
|---|---|---|---|
| a. | $4\frac{7}{8} + \underline{\phantom{xx}} = 5$ | b. | $7\frac{2}{5} + \underline{\phantom{xx}} = 8$ |
| c. | $3 = 2\frac{1}{6} + \underline{\phantom{xx}}$ | d. | $12 = 11\frac{1}{12} + \underline{\phantom{xx}}$ |

3. Use a number bond and the arrow way to show how to make one. Solve.

   a. $2\frac{3}{4} + \frac{2}{4}$
   
   $\frac{1}{4}$   $\frac{1}{4}$

   b. $3\frac{3}{5} + \frac{3}{5}$

Lesson 26:  Add a mixed number and a fraction.

4. Solve.

| | | | |
|---|---|---|---|
| a. | $4\frac{2}{3} + \frac{2}{3}$ | b. | $3\frac{3}{5} + \frac{4}{5}$ |
| c. | $5\frac{4}{6} + \frac{5}{6}$ | d. | $\frac{7}{8} + 6\frac{4}{8}$ |
| e. | $\frac{7}{10} + 7\frac{9}{10}$ | f. | $9\frac{7}{12} + \frac{11}{12}$ |
| g. | $2\frac{70}{100} + \frac{87}{100}$ | h. | $\frac{50}{100} + 16\frac{78}{100}$ |

Lesson 26: Add a mixed number and a fraction.

5. To solve $7\frac{9}{10} + \frac{5}{10}$, Maria thought, "$7\frac{9}{10} + \frac{1}{10} = 8$ and $8 + \frac{4}{10} = 8\frac{4}{10}$."

Paul thought, "$7\frac{9}{10} + \frac{5}{10} = 7\frac{14}{10} = 7 + \frac{10}{10} + \frac{4}{10} = 8\frac{4}{10}$." Explain why Maria and Paul are both right.

Name _____ Date _____

Solve.

1. $3\frac{2}{5} + $ _____ $= 4$

2. $2\frac{3}{8} + \frac{7}{8}$

Marta has 2 meters 80 centimeters of cotton cloth and 3 meters 87 centimeters of linen cloth. What is the total length of both pieces of cloth?

**Read**     **Draw**     **Write**

Lesson 27: Add mixed numbers.

Name _____ Date _____

1. Solve.

   a. $3\frac{1}{3} + 2\frac{2}{3} = 5 + \frac{3}{3} =$

   (number bonds: 3 and $\frac{1}{3}$; 2 and $\frac{2}{3}$)

   b. $4\frac{1}{4} + 3\frac{2}{4}$

   c. $2\frac{2}{6} + 6\frac{4}{6}$

2. Solve. Use a number line to show your work.

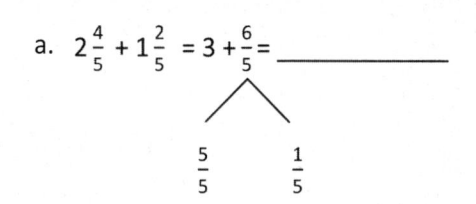

   a. $2\frac{4}{5} + 1\frac{2}{5} = 3 + \frac{6}{5} =$ _____

   (number bond: $\frac{5}{5}$ and $\frac{1}{5}$)

   b. $1\frac{3}{4} + 3\frac{3}{4}$

   c. $3\frac{3}{8} + 2\frac{6}{8}$

Lesson 27: Add mixed numbers.

3. Solve. Use the arrow way to show how to make one.

   a. $2\frac{4}{6} + 1\frac{5}{6} = 3\frac{4}{6} + \frac{5}{6} =$
   
   $\frac{5}{6}$ is split into $\frac{2}{6}$ and $\frac{3}{6}$

   b. $1\frac{3}{4} + 3\frac{3}{4}$

   c. $3\frac{3}{8} + 2\frac{6}{8}$

4. Solve. Use whichever method you prefer.

   a. $1\frac{3}{5} + 3\frac{4}{5}$

   b. $2\frac{6}{8} + 3\frac{7}{8}$

   c. $3\frac{8}{12} + 2\frac{7}{12}$

Lesson 27: Add mixed numbers.

Name _____  Date _____

Solve.

1. $2\frac{3}{8} + 1\frac{5}{8}$

2. $3\frac{4}{5} + 2\frac{3}{5}$

Matt had 2 m 65 cm of ribbon. He used 87 cm of the ribbon. How much ribbon did he have left?

**Read**     **Draw**     **Write**

Lesson 28: Subtract a fraction from a mixed number.

Name _____  Date _____

1. Subtract. Model with a number line or the arrow way.

   a. $3\frac{3}{4} - \frac{1}{4}$

   b. $4\frac{7}{10} - \frac{3}{10}$

   c. $5\frac{1}{3} - \frac{2}{3}$

   d. $9\frac{3}{5} - \frac{4}{5}$

2. Use decomposition to subtract the fractions. Model with a number line or the arrow way.

   a. $5\frac{3}{5} - \frac{4}{5}$

   (decomposition: $\frac{4}{5} = \frac{3}{5} + \frac{1}{5}$)

   b. $4\frac{1}{4} - \frac{2}{4}$

   c. $5\frac{1}{3} - \frac{2}{3}$

   d. $2\frac{3}{8} - \frac{5}{8}$

Lesson 28: Subtract a fraction from a mixed number.

3. Decompose the total to subtract the fractions.

a. $3\frac{1}{8} - \frac{3}{8} = 2\frac{1}{8} + \frac{5}{8} = 2\frac{6}{8}$

$2\frac{1}{8} \quad 1$

b. $5\frac{1}{8} - \frac{7}{8}$

c. $5\frac{3}{5} - \frac{4}{5}$

d. $5\frac{4}{6} - \frac{5}{6}$

e. $6\frac{4}{12} - \frac{7}{12}$

f. $9\frac{1}{8} - \frac{5}{8}$

g. $7\frac{1}{6} - \frac{5}{6}$

h. $8\frac{3}{10} - \frac{4}{10}$

i. $12\frac{3}{5} - \frac{4}{5}$

j. $11\frac{2}{6} - \frac{5}{6}$

Name _____ Date _____

Solve.

1. $10\frac{5}{6} - \frac{4}{6}$

2. $8\frac{3}{8} - \frac{6}{8}$

Jeannie's pumpkin had a weight of 3 kg 250 g in August and 4 kg 125 g in October. What was the difference in weight from August to October?

**Read**     **Draw**     **Write**

Lesson 29: Subtract a mixed number from a mixed number.

Name _____  Date _____

1. Write a related addition sentence. Subtract by counting on. Use a number line or the arrow way to help. The first one has been partially done for you.

   a. $3\frac{1}{3} - 1\frac{2}{3} =$ _____

   $1\frac{2}{3} +$ _____ $= 3\frac{1}{3}$

   b. $5\frac{1}{4} - 2\frac{3}{4} =$ _____

2. Subtract, as shown in Problem 2(a), by decomposing the fractional part of the number you are subtracting. Use a number line or the arrow way to help you.

   a. $3\frac{1}{4} - 1\frac{3}{4} = 2\frac{1}{4} - \frac{3}{4} = 1\frac{2}{4}$

   $\frac{1}{4} \quad \frac{2}{4}$

   b. $4\frac{1}{5} - 2\frac{4}{5}$

   c. $5\frac{3}{7} - 3\frac{6}{7}$

Lesson 29: Subtract a mixed number from a mixed number.

3. Subtract, as shown in Problem 3(a), by decomposing to take one out.

   a. $5\frac{3}{5} - 2\frac{4}{5} = 3\frac{3}{5} - \frac{4}{5}$
   
   (bond: $2\frac{3}{5}$ and $1$)

   b. $4\frac{3}{6} - 3\frac{5}{6}$

   c. $8\frac{3}{10} - 2\frac{7}{10}$

4. Solve using any method.

   a. $6\frac{1}{4} - 3\frac{3}{4}$  b. $5\frac{1}{8} - 2\frac{7}{8}$

   c. $8\frac{3}{12} - 3\frac{8}{12}$  d. $5\frac{1}{100} - 2\frac{97}{100}$

Lesson 29: Subtract a mixed number from a mixed number.

Name _____  Date _____

Solve using any strategy.

1. $4\frac{2}{3} - 2\frac{1}{3}$

2. $12\frac{5}{8} - 8\frac{7}{8}$

Lesson 30 Application Problem 4•5

There were $4\frac{1}{8}$ pizzas. Benny took $\frac{2}{8}$ of a pizza. How many pizzas are left?

**Read**     **Draw**     **Write**

Lesson 30: Subtract mixed numbers.

Name _____   Date _____

1. Subtract.

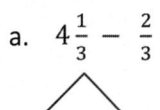

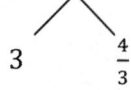

    a.  $4\frac{1}{3} - \frac{2}{3}$

        3    $\frac{4}{3}$

    b.  $5\frac{2}{4} - \frac{3}{4}$

    c.  $8\frac{3}{5} - \frac{4}{5}$

2. Subtract the ones first.

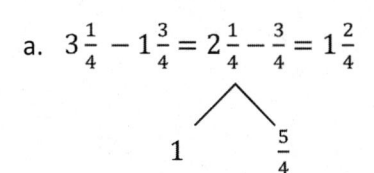

    a.  $3\frac{1}{4} - 1\frac{3}{4} = 2\frac{1}{4} - \frac{3}{4} = 1\frac{2}{4}$

        1    $\frac{5}{4}$

    b.  $4\frac{2}{5} - 1\frac{3}{5}$

**Lesson 30:** Subtract mixed numbers.

c. $5\frac{2}{6} - 3\frac{5}{6}$

d. $9\frac{3}{5} - 2\frac{4}{5}$

3. Solve using any strategy.

   a. $7\frac{3}{8} - 2\frac{5}{8}$

   b. $6\frac{4}{10} - 3\frac{8}{10}$

   c. $8\frac{3}{12} - 3\frac{8}{12}$

   d. $14\frac{2}{50} - 6\frac{43}{50}$

Lesson 30: Subtract mixed numbers.

Name _____ Date _____

Solve.

1. $7\frac{1}{6} - 2\frac{4}{6}$

2. $12\frac{5}{8} - 3\frac{7}{8}$

# A STORY OF UNITS – TEKS EDITION

Lesson 31 Problem Set 4•5

Name _____ Date _____

Use the RDW process to solve.

1. Tameka ran $2\frac{5}{8}$ miles. Her sister ran twice as far. How far did Tameka's sister run?

2. Natasha's sculpture was $5\frac{3}{16}$ inches tall. Maya's was 4 times as tall. How much shorter was Natasha's sculpture than Maya's?

3. A seamstress needs $1\frac{5}{8}$ yards of fabric to make a child's dress. She needs 3 times as much fabric to make a woman's dress. How many yards of fabric does she need for both dresses?

Lesson 31: Solve multiplicative comparison word problems involving fractions.

4. A piece of blue yarn is $5\frac{2}{3}$ yards long. A piece of pink yarn is 5 times as long as the blue yarn. Bailey tied them together with a knot that used $\frac{1}{3}$ yard from each piece of yarn. What is the total length of the yarn tied together?

5. A truck driver drove $35\frac{2}{10}$ miles before he stopped for breakfast. He then drove 5 times as far before he stopped for lunch. How far did he drive that day before his lunch break?

6. Mr. Washington's motorcycle needs $5\frac{5}{10}$ gallons of gas to fill the tank. His van needs 5 times as much gas to fill it. If Mr. Washington pays $3 per gallon for gas, how much will it cost him to fill both the motorcycle and the van?

Name _____ Date _____

Use the RDW process to solve.

Jeff has ten packages that he wants to mail. Nine identical packages weigh $2\frac{7}{8}$ pounds each. A tenth package weighs two times as much as one of the other packages. How many pounds do all ten packages weigh?